NUMERICAL METHODS USING Matlab

04

To our wives, Zena Lindfield and Wendy Penny, for their patience and support

NUMERICAL METHODS
USING MATLAB

Dr John Penny
George Lindfield
Department of Mechanical Engineering, Aston University

ELLIS HORWOOD
NEW YORK   LONDON   TORONTO   SYDNEY   TOKYO   SINGAPORE

First published 1995 by
Ellis Horwood Limited
Campus 400, Maylands Avenue
Hemel Hempstead
Hertfordshire, HP2 7EZ
A division of
Simon & Schuster International Group

Printed and bound in Great Britain by
Hartnolls Limited, Bodmin, Cornwall.

Library of Congress Cataloging-in-Publication Data

Penny, J. E. T. (John E. T.)
    Numerical methods using MATLAB / John Penny, George Lindfield.
        p.   cm.
    Includes bibliographical references and index.
    ISBN 013–030966–4 (pbk.)
    1. MATLAB.   2. Numerical analysis–Data processing.
I. Lindfield, G. R. (George R.)   II. Title.
QA297.P45   1995
519.4'0285'53–dc20   94–33327
                                                                        CIP

British Library Cataloguing in Publication Data

A catalogue record for this book is available
from the British Library

ISBN 0–13–030966–4

2 3 4 5   99 98 97 96 95

MATLAB is a registered trademark of The Mathworks, Inc., 24 Prime Park Way,
Natick, MA, USA, 01760–1500. Tel +(508) 653–1415, FAX +(508) 653–2997,
e-mail: info@mathworks.com.

Other trademarks mentioned in the text include; Control–C, a registered trademark of
SCT, Inc., Derive, a registered trademark of Soft Warehouse, Inc., IBM PC, a
trademark of International Business Machines, Inc., Macintosh, a trademark of Apple
Computers, Inc., Maple, a registered trademark of Waterloo Maple Software,
Mathematica, a trademark of Wolfram Research, Inc., MS Windows, a trademark of
Microsoft, Inc., NAg, a registered trademark of NAG Ltd., Sparc and Sun, trademarks
of Sun Microsystems, Inc., X Window, a trademark of Massachusetts Institute of
Technology

Contents

Contents

Contents

Contents

Preface

Our primary aim is to introduce the reader to a wide range of numerical algorithms, explain their fundamental principles and illustrate their application. The algorithms are implemented in the software package MATLAB® because it provides a powerful tool to help with these studies.

Many important theoretical results are discussed but it is not intended to provide a detailed and rigorous theoretical development in every area. We wish to show how numerical procedures can be applied to solve a range of problems and how the algorithms, when applied to particular problems, demonstrate the expected theoretical performance.

When used with care MATLAB provides a natural and succinct way of describing numerical algorithms and a powerful means of experimenting with them. Often MATLAB allows a remarkably direct translation from the statement of the algorithm to its expression as a script. However, no tool, irrespective of its power, should be used carelessly or uncritically.

This text should allow the reader to study numerical methods by encouraging systematic experimentation with some of the many fascinating problems of numerical analysis. The text introduces the reader to a range of useful and important algorithms and develops MATLAB functions to implement them. The reader is encouraged to use these functions to produce results in numerical and graphical form. Particular examples are given throughout the book to illustrate how numerical methods are used to study problems which include applications in the biosciences, chaos, neural networks, engineering and science.

Preface

It should be noted that the introduction to MATLAB is relatively brief and is meant only as an aid to the reader. It can in no way be expected to replace the standard MATLAB manual. A particular feature of the book is that it provides supplemental functions and scripts developed with MATLAB, v4. Thus we are able use the new facilities provided for sparse matrices and the greatly enhanced graphics to introduce the reader to the more difficult topics. However, apart from those functions and scripts which use the most recent features of MATLAB v4 the rest are compatible with earlier versions of MATLAB. The scripts and functions we provide are, as far as we are aware, platform independent. They have been kept as simple as possible for reasons of clarity. They could be improved and we urge readers to develop the ones that are of particular interest to them. These supplemental functions and scripts are available from The MathWorks, Inc. free of charge on PC or Macintosh diskette. To request a copy please complete and return the card bound in the back of this book.

The text provides a general introduction to MATLAB; the solution of linear equations and eigenvalue problems; methods for solving non-linear equations; numerical integration and differentiation; the solution of initial value and boundary value problems; curve fitting including splines, least squares and Fourier analysis. The last chapter presents topics in optimisation such as interior point methods, non-linear programming and genetic algorithms which we hope will attract the interest of readers. We provide a broad introduction to the topics and encourage the reader to experiment with the MATLAB functions provided. The text also provides some MATLAB execution times which, in general, have been produced using a Macintosh LC II, but a table providing a comparison with other machines is given in Chapter 1.

The text contains many worked examples, practice problems and solutions and we hope we have provided an interesting range of problems. The text is suitable for undergraduate and postgraduate students and for those working in industry and education. We hope readers will share our enthusiasm for this area of study. For those who do not currently have access to MATLAB, this text still provides a general introduction to a wide range of numerical algorithms and many useful examples.

We would be pleased to hear from readers who note errors or have suggestions for improvements. We would also like to thank Karen Rose, Jacqueline Harbor of Prentice Hall and Cristina Palumbo of The Mathworks, Inc. for their help and encouragement in producing this text.

George Lindfield and John Penny
Aston University
Birmingham
July 1994

1

An introduction to MATLAB

1.1 THE SOFTWARE PACKAGE MATLAB

MATLAB® is a software package produced by The MathWorks, Inc. and is available on systems ranging from personal computers to the Cray super-computer. The name MATLAB is derived from the term "matrix laboratory". It provides an interactive development tool for scientific and engineering problems and more generally for those areas where significant numeric computations have to be performed. The package can be used to evaluate single statements directly or a list of statements called a script can be prepared. Once named and saved, a script can be executed as an entity in much the same way as a program. The package is based on software produced by the LINPACK and EISPACK projects which represent the "state-of-the-art" numerical software for matrix computations. MATLAB provides the user with:

(1) easy manipulation of matrix structures,
(2) a vast number of powerful inbuilt routines which is constantly growing and developing,
(3) powerful two- and three-dimensional graphing facilities,
(4) a scripting system which allows users to develop and modify the software for their own needs.

It is not difficult to use MATLAB, although to use it with maximum efficiency requires experience. Essentially MATLAB works with rectangular or square arrays of data (matrices), the elements of which may be real or complex. A scalar quantity is thus a matrix containing a single element. This is an elegant and powerful notion but it can present the user with an initial conceptual difficulty. A user schooled in such languages as FORTRAN, BASIC or Pascal is familiar with a pseudo-statement of the form A = abs(B) and will immediately interpret it as an instruction that A is assigned the absolute value of the number stored in B. In MATLAB the variable B may represent a matrix so that *each element* of the matrix A will become the absolute value of the corresponding element in B.

There are several software packages that have some similarities to MATLAB. These packages include:

The NAg Library. This is a very extensive, high-quality collection of subroutines for numerical analysis but some users may find it more difficult to use than MATLAB since the subroutines must be called from a programming language such as FORTRAN, BASIC or Pascal.

Mathematica, Maple and *Derive.* These packages place a major emphasis on symbolic mathematical manipulation. There is now a MATLAB symbolic manipulation toolbox. It consists of the embedded Maple V library and over 50 M-files which provide MATLAB versions of Maple functions for such tasks as calculus, symbolic equation solving and variable precision arithmetic. In addition there is an extended symbolic manipulation toolbox with the complete Maple V library of functions.

APL. The letters stand for A Programming Language. This is a language designed mainly for manipulating matrices. It contains many powerful facilities but the symbols it uses are non-standard and the syntax unusual. The keyboard must be remapped for these special characters.

Control-C and *Octave.* These packages are also developed from the LINPACK and EISPACK projects. Some of the commands and functions are very similar to those of MATLAB.

1.2 MATLAB ON PERSONAL COMPUTERS AND WORKSTATIONS

The current MATLAB release, version 4, is available on a wide variety of platforms and functions under a variety of operating systems. In particular it is implemented on the Sun Sparc workstation under X Windows, on IBM PC compatibles under MS Windows and on Macintosh computers. When MATLAB is invoked it opens a command window, and if required graphics, editing and help windows may also be opened. MATLAB scripts and function are in general platform independent and they can be readily ported from one system to another. To install and start MATLAB readers should consult the manual appropriate to their particular working environment.

The scripts and functions given in this book have been tested under MATLAB version 4. However, most of them will work directly using earlier versions of MATLAB but some will require modification, particularly those involving graphics and sparse matrices.

The remainder of this chapter is devoted to introducing some of the statements and syntax of MATLAB. The intention is to give the reader a sound but brief introduction to the power of MATLAB. Some details of structure and syntax will be omitted and must be obtained from the MATLAB manual.

1.3 MATRICES AND MATRIX OPERATIONS IN MATLAB

The matrix is fundamental to MATLAB and we have provided a broad and simple introduction to matrices in Appendix 1. In MATLAB the names used for matrices must start with a letter and may be followed by any combination of letters or digits. The letters may be upper or lower case. Note that throughout this text a distinctive font is used to denote MATLAB statements and output.

In MATLAB the arithmetic operations of addition, subtraction, multiplication and division can be performed directly with matrices and we will now examine these commands. First of all the matrices must be created. There are several ways of doing this in MATLAB and the simplest method, which is suitable for small matrices, is as follows. We will assign an array of values to A by opening the **command** window and then typing

```
A=[1 3 5;1 0 1;5 0 9]
```

When the return key is pressed the matrix will be displayed thus:

```
A =
    1    3    5
    1    0    1
    5    0    9
```

By typing, for example, B=[1 3 51;2 6 12;10 7 28] and pressing the return key we assign values to B. All statements are terminated by pressing return. To add the matrices in the **command** window and assign the result to C we type C=A+B and similarly if we type C=A-B the matrices will be subtracted. In both cases the result will be printed row by row. Note that terminating a MATLAB statement with a semicolon suppresses the output. If these statements are typed in an **edit** window as part of a script (i.e. a program) then there will be no execution or output until the script is run by typing its name in the **command** window and pressing return.

The implementation of vector and matrix multiplication in MATLAB is straightforward. Beginning with vector multiplication, we will assume that row vectors having the same number of elements have been assigned to d and p. To multiply them together we write x=d*p'. Note that the symbol ' transposes the row p into a column so that the multiplication is valid. The result, x, is a scalar. For matrix multiplication, assuming the two matrices A and B have been assigned, the user simply types C=A*B. This will compute A postmultiplied by B, assign the result to C and display it if the multiplication is valid; otherwise MATLAB will give an appropriate error indication. The conditions for matrix multiplication to be valid are given in Appendix 1. Notice that the symbol * must be used for multiplication because in MATLAB multiplication is not implied.

1.4 USING THE MATLAB OPERATOR \ FOR MATRIX DIVISION

It is easy to solve the problem $ax = b$ where a and b are simple scalar constants and x is the unknown. Given a and b then $x = b/a$. However, consider the corresponding matrix equation

$$\mathbf{Ax} = \mathbf{b} \tag{1.4.1}$$

where **A** is a square matrix. We now wish to find **x** where **x** and **b** are vectors. Computationally this is a much more difficult problem and in MATLAB it is solved by executing the statement

    x = A\b

This statement uses the important MATLAB division operator \ and solves the linear equation system (1.4.1).

Solving linear equation systems is an important problem and the computational efficiency and other aspects of this type of problem are discussed in considerable detail in Chapter 2.

1.5 MANIPULATING THE ELEMENTS OF A MATRIX

In MATLAB matrix elements can be manipulated individually or in blocks. For example, X(1,3)=C(4,5)+V(9,1), a(1)=b(1)+d(1) or C(i,j+1)=D(i,j+1)+E(i,j+1) are valid statements relating elements of matrices. Rows and columns can be manipulated as complete entities. Thus A(:,3), A(5,:) refer respectively to the third column and fifth row of A. If B is a 10 by 10 matrix then B(:,4:9) refers to columns 4 to 9 of the matrix. Note that in MATLAB, by default, the *lowest matrix index starts at 1*.

The following examples illustrate some of the ways subscripts can be used in MATLAB. First we assign a matrix

    »a=[2 3 4 5 6;-4 -5 -6 -7 -8;3 5 7 9 1;4 6 8 10 12;-2 -3 -4 -5 -6]

    a =
         2   3   4   5   6
        -4   -5   -6   -7   -8
         3   5   7   9   1
         4   6   8   10   12
        -2   -3   -4   -5   -6

Executing the following statements

    »v=[1 3 5];
    »b=a(v,2)

gives

```
b =
     3
     5
    -3
```

Thus b is composed of the elements of the first, third and fifth rows in the second column of a. Executing

```
»c=a(v,:)
```

gives

```
c =
     2      3      4      5      6
     3      5      7      9      1
    -2     -3     -4     -5     -6
```

Thus c is composed of the first, third and fifth rows of a. Executing

```
»d=zeros(3)
»d(:,1)=a(v,2)
```

gives

```
d =
     3      0      0
     5      0      0
    -3      0      0
```

Here d is a 3 x 3 matrix of zeros with column 1 replaced by the first, third and fifth elements of column 2 of a. Executing

```
»e=a(1:2,4:5)
```

gives

```
e =
     5      6
    -7     -8
```

1.6 TRANSPOSING MATRICES

A simple operation that may be performed on a matrix is transposition which interchanges rows and columns. In MATLAB this is denoted by the symbol '. For example:

```
»a=[1 2 3;4 5 6;7 8 9]

a =
     1      2      3
     4      5      6
     7      8      9
```

To assign the transpose of *a* to *b* we write

```
»b=a'

b =
     1     4     7
     2     5     8
     3     6     9
```

However, if *a* is complex then the MATLAB operator ' gives the complex conjugate transpose. For example:

```
»a=[1+2*i 3+5*i;4+2*i 3+4*i]

a =
   1.0000 + 2.0000i   3.0000 + 5.0000i
   4.0000 + 2.0000i   3.0000 + 4.0000i

»b=a'

b =
   1.0000 - 2.0000i   4.0000 - 2.0000i
   3.0000 - 5.0000i   3.0000 - 4.0000i
```

To provide the transpose without conjugation, we execute

```
»b=a.'

b =
   1.0000 + 2.0000i   4.0000 + 2.0000i
   3.0000 + 5.0000i   3.0000 + 4.0000i
```

1.7 SPECIAL MATRICES

Certain matrices occur frequently in matrix manipulations and MATLAB ensures that these are generated easily. Some of the most common are ones(m,n), zeros(m,n), rand(m,n) and randn(m,n). These MATLAB statements generate *m* x *n* matrices composed of ones, zeros, randomly generated and normally randomly generated elements respectively. The MATLAB statement eye(n) generates the *n* x *n* unit matrix. If only a single scalar parameter is given, then these statements generate a square matrix of the size given by the parameter. If we wish to generate an identity matrix B of the same size as an already existing matrix A, then the statement B=eye(size(A)) can be used. Similarly C=zeros(size(A)) and D=ones(size(A)) will generate a matrix C of zeros and a matrix D of ones, both of which are the same size as matrix A.

1.8 GENERATING MATRICES WITH SPECIFIED ELEMENT VALUES

Here we will confine ourselves to some relatively simple examples thus:

x=-10:1:10 sets x to a vector having elements –10, –9, –8, ..., 8, 9, 10.
y=-2:.2:2 sets y to a vector having elements –2, –1.8, –1.6, ..., 1.8, 2.
z=[1:3 4:2:8 10:0.5:11] sets z to a vector having the elements

[1 2 3 4 6 8 10 10.5 11]

More complex matrices can be generated from others. For example, consider the two statements

```
C=[2.3 4.9; 0.9 3.1];
D=[C ones(size(C));eye(size(C)) zeros(size(C))]
```

These two statements generate a new matrix D the size of which is double that of the original C. The matrix will have the form

$$D = \begin{bmatrix} 2.3 & 4.9 & 1 & 1 \\ 0.9 & 3.1 & 1 & 1 \\ 1 & 0 & 0 & 0 \\ 0 & 1 & 0 & 0 \end{bmatrix}$$

1.9 SOME SPECIAL MATRIX OPERATIONS

Some arithmetic operations are simple to execute for single scalar values but involve a great deal of computation for matrices. For large matrices such operations may take a significant amount of time. An example of this is where a matrix is raised to a power. We can write this in MATLAB as A^p where p is a positive scalar value and A is a matrix. This produces the power of the matrix and may be obtained in MATLAB for any value of p. For the case where the power equals 0.5 it is better to use sqrtm(A) which gives the square root of the matrix A. Another special operation directly available in MATLAB is expm(A) which gives the exponential of the matrix A. Other special operators are available in MATLAB which involve operations between the individual elements of a matrix. These are discussed in the following section.

1.10 ELEMENT-BY-ELEMENT OPERATIONS

Element-by-element operations differ from the standard matrix operations but they can be very useful. They are achieved by using a period or dot (.) to precede the operator. For example, X.^Y, X.*Y and Y.\X. If, in these statements, X and Y are matrices (or vectors) the *elements* of X are raised to the power, multiplied o by the *corresponding element* of Y depending on the operator used. The f gives the same result as the division operation specified above. For these o to be executed the matrices and vectors used must be the same size. Note tha is *not* used in the operations + and – because ordinary matrix addition and s

Combinations of strings can be printed using square brackets [ ] and numerical values can be placed in text strings if they are converted to strings using the num2str function. For example,

```
x=2.678;
disp(['Value of iterate is ', num2str(x),' at this stage']);
```

will place on the screen

```
Value of iterate is  2.678 at this stage
```

The more flexible fprintf function allows formatted output to the screen or to a file. It takes the form

```
fprintf('filename','format string',list);
```

where list is a list of variable names separated by commas. The filename parameter is optional; if not present output is to the screen. The format string formats the output. The elements that may be used in the format string are

> %P.Qe for exponential notation
> %P.Qf fixed point
> %P.Qg becomes %P.Qe or %P.Qf whichever is shorter
> /n gives new line

where P and Q, above, are integers. The integer string characters, including a period (.), must follow the % symbol and precede the letter. The integer before the period sets the field width; the integer after the period sets the number of decimal places after the decimal point. For example, %8.4f and %10.3f give field width 8 with four decimal places and 10 with three decimal places respectively. Note that one space is allocated to the decimal point. For example,

```
x=1007.46; y=2.1278; k=17;
fprintf('/nx= %8.2f y= %8.6f k= %2.0f/n',x,y,k);
```

outputs

```
x=1007.46 y=  2.127800 k=17
```

We now consider the input of text and data from the keyboard. An interactive way of obtaining input is to use the function input. This takes the forms

```
Variable=input('text');  or Variable=input('text','s');
```

The first form displays a prompt text and allows the input of *numerical values*. Single values or matrices can be entered in this way. The input function displays the text as a prompt and then waits for an entry. This is assigned to the variable when return is pressed. The second form allows for the entry of *strings*.

For large amounts of data, perhaps saved in a previous MATLAB session, the function load allows the loading of files from disk using

    load filename

1.12 MATLAB GRAPHICS

MATLAB provides a wide range of graphics facilities which may be called from within a script or used simply in command mode for direct execution. We will begin by considering the plot function. This function takes several forms. For example:

> plot(x,y)   plots the vector x against y. If x and y are matrices the first column of x is plotted against the first column of y. This is then repeated for each pair of columns of x and y.

> plot(x1,y1,'linetype_or_pointtype1',x2,y2,'linetype_or_pointtype2') plots the vector x1 against y1 using the linetype_or_pointtype1; then the vector x2 against y2 using the linetype_or_pointtype2.

The linetype_or_pointtype is selected by using the required symbol from the following table:

Lines	Symbol	Points	Symbol
solid	–	point	.
dashed	–	plus	+
dotted	:	star	*
dashdot	-.	circle	o
		x mark	x

Semilog and loglog graphs can be obtained by replacing plot by semilog or loglog and various other replacements for plot are available to give special plots.

Titles, axis labels and other features can be added to a given graph using the functions xlabel, ylabel, title, grid and text. These functions have the following form:

> title('title') displays title at top of graph.

> xlabel('x_axis_name') displays name chosen for x_axis.

> ylabel('y_axis_name') displays name chosen for y_axis.

> grid superimposes a grid on the graph.

> text(x,y,'text-at-x,y') displays text-at-x,y at position (x, y) in the **graph** window where x and y are measured in the units of the current plotting axes. There may be one point or many at which text is placed depending on whether or not x and y are vectors.

gtext('text') allows the placement of text using the mouse by positioning it where the text is required and then pressing the button.

In addition, the function axis allows the user to set the limits of the axes for a particular plot. This takes the form axis(p) where p is a four-element row vector specifying the lower and upper limits of the axes in the x and y directions. The axis statement must be placed after the plot statement to which it refers. Note that the functions title, xlabel, ylabel, grid, text, gtext and axis must *follow* the plot to which they refer.

The following script gives a plot which is output as Fig. 1.12.1. The function hold is used to ensure that the two graphs are superimposed.

```
x=-4:0.05:4; y=exp(-0.5*x).*sin(5*x);
figure(1);
plot(x,y);
xlabel('x-axis'); ylabel('y-axis');
hold on;
y=exp(-0.5*x).*cos(5*x);
plot(x,y);
grid; gtext('Two tails...');
hold off
```

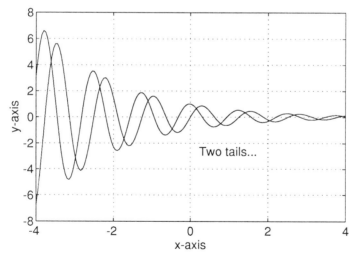

Fig. 1.12.1. Superimposed graphs obtained using plot(x,y) and hold statements.

The function fplot allows the user to plot a previously defined function between given limits. The important difference between fplot and plot is that fplot chooses the plotting points in the given range adaptively. Thus more points are chosen when the function is changing more rapidly. This is illustrated by executing the following MATLAB script:

```
x=2:.04:4;
y=f101(x);
plot(x,y);
xlabel('x');  ylabel('y');
figure(2)
fplot('f101',[2 4],10)
xlabel('x');  ylabel('y');
```

The function f101 is given by

```
function v=f101(x)
v=sin(x.^3);
```

Running the above script produces Fig. 1.12.2 and Fig. 1.12.3. In this example we have deliberately chosen an inadequate number of plotting points but even so function fplot has produced a smoother and more accurate curve.

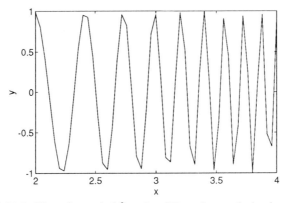

Fig. 1.12.2. Plot of $y = \sin(x^3)$ using 75 equispaced plotting points.

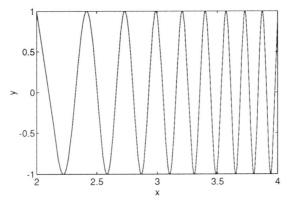

Fig. 1.12.3. Plot of $y = \sin(x^3)$ using the function fplot to choose plotting points adaptively.

There are a number of special features available in MATLAB for the presentation and manipulation of graphs and some of these will now be discussed. The subplot function takes the form subplot(p,q,r) where p, q splits the figure window into a *p* by *q* grid of cells and places the plot in the *r*th cell of the grid, numbered consecutively along the rows. This is illustrated by running the following script which generates six different plots, one in each of the six cells. These plots are given in Fig. 1.12.4.

```
x=0.1:.1:5;
subplot(2,3,1);plot(x,x);title('plot of x');
xlabel('x'); ylabel('y');
subplot(2,3,2);plot(x,x.^2);title('plot of x^2');
xlabel('x'); ylabel('y');
subplot(2,3,3),plot(x,x.^3);title('plot of x^3');
xlabel('x'); ylabel('y');
subplot(2,3,4),plot(x,cos(x));title('plot of cos(x)');
xlabel('x'); ylabel('y');
subplot(2,3,5),plot(x,cos(2*x));title('plot of cos(2x)');
xlabel('x'); ylabel('y');
subplot(2,3,6),plot(x,cos(3*x));title('plot of cos(3x)');
xlabel('x'); ylabel('y');
```

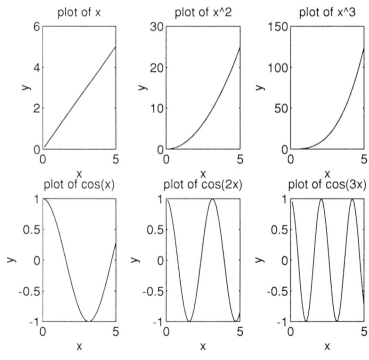

Fig. 1.12.4. Example of the use of subplot.

The current plot can be held on screen by using the function hold and subsequent plots are drawn over it. The function hold on switches the hold facility on while hold off switches it off. The figure window can be cleared using the function clf.

Information can be taken from a graphics window using the function ginput which takes two main forms. The simplest is

[x,y]=ginput

This inputs an unlimited number of points to the vectors x and y by positioning the mouse cross-hairs at the points required and then pressing the mouse button. To exit ginput the return key must be pressed. If a specific number of points n are required then we write

[x,y]=ginput(n)

1.13 THREE-DIMENSIONAL GRAPHICS

It is often convenient to draw a three-dimensional graph of a function or set of data to gain a deeper insight into the nature of that data. MATLAB provides powerful and extensive facilities to allow the user to draw a wide range of three-dimensional graphs. Here we only briefly introduce a small selection of the these functions. These are the functions meshgrid, mesh, surfl, contour and contour3. It should be noted that the more complex graphs of this type may take a significant time to draw on the screen, depending on the algebraic complexity of the function, the amount of detail required and the power of the computer being used. Consequently the user should sometimes be prepared for a significant wait.

Usually three-dimensional functions are plotted to illustrate particular features of the function such as regions where maxima or minima lie. Plotting surfaces to illustrate these features can be difficult and some careful analysis of the function may be needed before the graph is drawn successfully. In addition, even when the region of interest is successfully located and plotted the feature of interest may be hidden and it will then be necessary to choose a different viewpoint. Discontinuities may also be present and cause plotting problems.

For the function $z = f(x, y)$ the MATLAB function meshgrid is used to *generate a complete set of points in the x–y plane for the three-dimensional plotting functions.* We can then compute the values of z and these are finally plotted by using one of the functions mesh, surf, surfl or surfc. For example, to plot the function

$$z = (-20x^2 + x)/2 + (-15y^2 + 5y)/2 \text{ for } x = -4{:}0.2{:}4 \text{ and } y = -4{:}0.2{:}4$$

we first set up the values of the x–y domain and then compute z corresponding to these x and y values using the given function. Finally we plot the three-dimensional graph using the function surfl. This is achieved by using the following script. Note how the function figure is used to direct the output to a graphics window so that the first plot is not over-written by the second.

```
clf
[x,y]=meshgrid(-4.0:0.2:4.0,-4.0:0.2:4.0);
z=0.5*(-20*x.^2+x)+0.5*(-15*y.^2+5.*y);
figure(1);
surfl(x,y,z);
axis([-4 4 -4 4 -400 0] )
xlabel('x-axis'); ylabel('y-axis'); zlabel('z-axis');
figure(2);
contour3(x,y,z,15);
axis([-4 4 -4 4 -400 0] )
xlabel('x-axis'); ylabel('y-axis'); zlabel('z-axis');
```

Running this script generates the plots shown in Fig. 1.13.1 and Fig. 1.13.2. The former is created using surfl and shows the function as a surface. The latter is created by contour3 and is a three-dimensional contour plot of the surface.

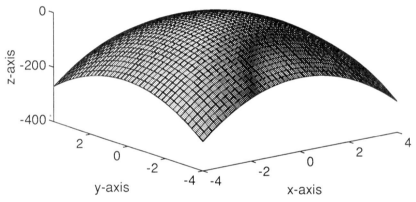

Fig. 1.13.1. Three-dimensional surface using default view.

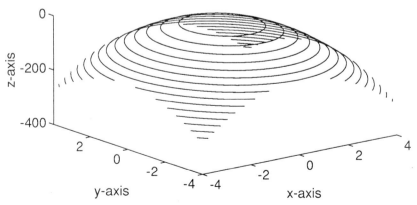

Fig. 1.13.2. Three-dimensional contour plot.

When plotting surfaces a very useful function is `view`. This function allows the surface or mesh to be viewed from different positions. The function has the form `view(az, el)` where `az` is the azimuth and `el` the elevation of the viewpoint required. Azimuth may be interpreted as the viewpoint rotation about the z-axis and elevation as the rotation of the viewpoint about the x–y plane. A positive value of the elevation gives a view from above the object and a negative value a view from below. Similarly a positive value of azimuth gives a counter-clockwise rotation of the viewpoint about the z-axis while a negative value gives a clockwise rotation. If the view function is not used the default values are –37.5° for the azimuth and 30° for the elevation.

There are many other plotting facilities but they are not described here.

1.14 SCRIPTING IN MATLAB

In some of the previous sections we have created some simple MATLAB scripts that have allowed a series of commands to be executed sequentially. However, many of the features usually found in programming languages are also provided in MATLAB to allow the user to create versatile scripts. The more important of these features will be described in this section. It must be noted that scripting is done in the **edit** window using a text editor appropriate to the system, not in the **command** window which only allows the execution of statements one at a time or several statements provided that they are on the same line.

MATLAB *does not require the declaration of variable types* but for the sake of clarity the role and nature of key variables may be indicated by using comments. Any text following the symbol % is considered as comment. In addition there are certain variable names which have predefined special values for the convenience of the user. They can, however, be redefined if required. These are

eps	which is set to the particular machine accuracy
pi	which equals π
inf	the result of dividing by zero
NaN	which is not a number produced on dividing zero by zero
i,j	both of which equal $\sqrt{-1}$.

Assignment statements in a MATLAB script take the form

```
variable=expression;
```

The expression is calculated and the value assigned to the variable on the left-hand side. If the semicolon is omitted from the end of these statements the name of the variable(s) and the assigned value(s) is displayed on the screen. A further form is not to assign the expression explicitly. In this case the value of the expression is calculated, assigned to the variable `ans` and displayed.

A variable in MATLAB is assumed to be a matrix of some kind; its name must start with a letter and may be followed by any combination of digits and letters; a maximum of 19 characters is recognised. It is good practice to use a meaningful variable name.

However, it is very important to avoid the use of existing MATLAB commands, function names or even the word MATLAB itself! MATLAB does not prevent their use but using them can lead to problems and inconsistencies. The expression is a valid combination of variables, constants, operators and functions. Brackets can be used as convenient to alter or clarify the precedence of operations. The precedence of operation for simple operators is first ^, second *, third / and finally + and - where

^   raises to a power
*   multiplies
/   divides
+   adds
-        subtracts

The effects of these operators in MATLAB have already been discussed.

Unless there are instructions to the contrary, a set of MATLAB statements in a script will be executed in sequence. This is the case in the following example.

```
% Matrix calculations for two matrices A and B
A=[1 2 3; 4 5 6;7 8 9];
B=[5 -6 -9;1 1 0; 24 1 0];
% Addition result assigned to C
C=A+B;
disp(C);
% Multiplication result assigned to D
D=A*B;
disp(D);
% Division result assigned to E
E=A\B;
disp(E);
```

In order to allow the repeated execution of one or more statements a for loop is used. This takes the form

```
for loopvariable = loopexpression
  Statements
end
```

The loopvariable is a suitably named variable and loopexpression is usually of the form n:m or m:i:n where n, i and m are the initial, incremental and final values of loopvariable. They may be constants, variables or expressions; they can be negative or positive but clearly they should be chosen to take values consistent with the logic of the script. This structure should be used when the loop is to be repeated a predetermined number of times.

Example:

```
for i=1:n
  for j=1:m
    C(i,j)=A(i,j)+cos((i+j)*pi/(n+m))*B(i,j);
  end
end

for k=n+2:-1:n/2
  a(k)=sin(pi*k);
  b(k)=cos(pi*k);
end
```

When the repetition is continued subject to a condition being satisfied which may be dependent on values generated within the loop the while statement is used. This has the form

```
while while_expression
  statements
end
```

The while_expression is a *relational expression* of the form e1°e2 where e1 and e2 are ordinary arithmetic expressions as described before and ° is a relational operator defined as follows:

==	equal
<=	less than or equal
>=	greater than or equal
~=	not equal
<	less than
>	greater than

Relational expressions may be combined using the following *logical operators*:

& provides the and operator; | provides the or operator;
~ provides the not operator.

Note that false is 0 and true is non-zero. Relational operators have a higher order of precedence than logical operators.

Examples of while loops:

```
dif=1;
while dif>0.0005
  x1=x2-cos(x2)/(1+x2);
  dif=abs(x2-x1);
  x2=x1;
end
```

```
while sum(x) ~= max(y)
   x=x.^2;
   y=y+x;
end
```

Note also that break allows exit from while or for loops.

A vital feature of all programming languages is the ability to change the sequence in which instructions are executed, subject to some condition being satisfied that may be dependent on values generated within the program. In MATLAB the if statement is used to achieve this and has the general form

```
if if_expression
   statements
elseif if_expression
   statements
elseif if_expression
   statements
      ...
      ...
else
   statements
end
```

Here the if_expression is a relational expression of the form e1°e2 where e1 and e2 are ordinary arithmetic expressions and ° is a relational operator as described before. Relational expressions may be combined using logical operators.

Examples:

```
for k=1:n
   for p=1:m
      if k==p
         z(k,p)=1;
         total=total+z(k,p);
      elseif k<p
         z(k,p)=-1;
         total=total+z(k,p);
      else
         z(k,p)=0;
      end
   end
end

if (x~=0)&(x<y)
   b=sqrt(y-x)/x;
   disp(b);
end
```

It is important to note that if any statement in a MATLAB script is longer than one line then it must be continued by using an ellipsis (...) at the end of a line to denote a continuation onto the next line.

1.15 FUNCTIONS IN MATLAB

A very large number of functions are directly built into MATLAB. They may be called by using the function name together with the parameters that define the function. These functions may return one or more values. A small selection of MATLAB functions are given in the following table which gives the function name, the function use and an example function call. Note that all function names must be in lower case letters.

Function	Function gives	Example
sqrt(x)	square root of x	y=sqrt(x+2.5);
abs(x)	positive value of x	d=abs(x)*y;
conj(x)	the complex conjugate of x	x=conj(y);
sin(x)	sine of x	t=x+sin(x);
log(x)	log to base e of x	z=log(1+x);
log10(x)	log to base 10 of x	z=log10(1-2*x);
cosh(x)	hyperbolic cosine of x	u=cosh(pi*x);
exp(x)	exponential of x, i.e. e^x	p=.7*exp(x);
gamma(x)	gamma function of x	f=gamma(y);
expm(A)	exponential of matrix **A**	p=expm(A+B);
sqrtm(A)	square root of matrix **A**	y=sqrtm(A)

In the case of the functions expm(A) and sqrtm(A) the matrix is treated as a whole. For example, B=sqrtm(A) assigns the square root of the matrix A to B so that the matrix B multiplied by itself is equal to A. In contrast, C=sqrt(A) takes the square root of each element of A and assigns them to C.

Some functions perform special calculations for an important and general mathematical process. These functions often require more than one input parameter and may provide several outputs. For example, bessel(n,x) gives the nth order Bessel function of x. The statement y=fzero('fun',x0) determines the root of the function fun near to x0 where fun is a function given by the user that defines the equation for which we are finding the root. The statement [Y,I]=sort(X) is an example of a function that can return two output values. Y is the sorted matrix and I is a matrix containing the indices of the sort.

In addition to a large number of mathematical functions, MATLAB provides several useful utility functions that may be used for examining the operation of scripts. These are:

clock   Returns the current date and time in the form: [year month day hour min sec].

etime(t2,t1)   Calculates elapsed time between t1 and t2. This is very useful for timing segments of a script.

tic,toc   An alternative way of finding the time taken to execute a segment of script. The statement tic starts the timing and toc gives the elapsed time since the last tic.

cputime   Returns the total time in seconds since MATLAB was launched.

flops   Allows the user to check the number of floating point operations, that is the total number of additions, subtractions, multiplications and divisions in a section of script. Executing flops(0) starts the count of flops at zero; flops counts the number of flops cumulatively.

pause   Causes the execution of the script to pause until the user presses a key. Note that the cursor is turned into the symbol P, warning the script is in pause mode. This is often used when the script is operating with echo on.

echo on   Displays each line of script in the **command** window before execution. This is useful for demonstrations. To turn it off use the statement echo off.

The following script uses the timing functions described above to estimate the time taken to solve a 100 x 100 system of linear equations:

```
%Solve 100x100 system
A=rand(100);b=rand(100,1);
Tbefore=clock; tic; t0=cputime; flops(0);y=A\b;
timetaken=etime(clock,Tbefore); tend=toc; t1=cputime-t0;
disp('etime    tic-toc    cputime ')
fprintf('%5.2f%10.2f%10.2f\n\n', timetaken,tend,t1);
count=flops;
fprintf('flops = %6.0f\n',count);
```

Running this script gives

```
»etime    tic-toc    cputime
 6.92      6.95       7.00

flops = 732005
```

The output shows that the three alternative methods of timing give essentially the same value. The flops are also counted and displayed.

1.16 USER-DEFINED FUNCTIONS

In our description of functions we have mentioned user-defined functions. MATLAB allows users to define their own functions but a specific form of definition must be followed. This is very simple and may be described as follows:

```
function  output_params = function_name(input_params)
function body
```

Once the function is defined *it must be saved as an M-file* so that it can then be used. It is good practice to put some comment describing the nature of the function after the function heading. This comment may then be accessed via the help command or menu.

The function call takes the form

```
specific_output_params = function_name(specific_input_params);
```

where specific_output_params is either one parameter name or, if there is more than one output parameter, a list of these parameters placed in square brackets, separated by commas. The specific_input_params term is either a single parameter or a list of parameters separated by commas. The function body consists of the statements defining the user's function. These statements *should include statements assigning values to the output parameters*. For example, suppose we wish to define the function

$$\left(\frac{x}{2.4}\right)^3 - \frac{2x}{2.4} + \cos\left(\frac{\pi x}{2.4}\right)$$

The MATLAB function definition will be as follows:

```
function p=fun1(x)
% A simple function definition
x=x/2.4;
p=x^3-2*x+cos(pi*x);
```

A call of this function is y=fun1(2.5); or z=fun1(x-2.7*y); assuming x and y have preset values. Another way of calling the function is as an input parameter to another function. For example:

```
solution=fzero('fun1',2.9);
```

This will give the zero of fun1 closest to 2.9.

The following example illustrates the definition of a function with more than one input and one output parameter:

```
function [x1,x2]=rootquad(a,b,c)
% This function solves a simple quadratic
% equation ax^2+bx+c =0 given the
% coefficients a,b,c. The solutions are
% assigned to x1 and x2
d=b*b-4*a*c;
x1=(-b+sqrt(d))/(2*a);
x2=(-b-sqrt(d))/(2*a);
```

This function solves quadratic equations and, assuming x, y and z have preset values, valid function calls are

are element-by-element operations. Examples of element-by-element operations are given below:

```
»a=[1 2;3 4]

a =
        1       2
        3       4

»b=[5 6;7 8]

b =
        5       6
        7       8

»a*b

ans =
       19      22
       43      50
```

This gives normal matrix multiplication. However, using the . operator we have

```
»a.*b

ans =
        5      12
       21      32
```

which is element-by-element multiplication.

```
»a.^b

ans =
            1              64
         2187           65536
```

In the above, each element of a is raised to the corresponding power in b.

1.11 INPUT AND OUTPUT IN MATLAB

To output the names and values of variables the semicolon can be omitted from assignment statements. However, this does not produce clear scripts or well-organised and tidy output. It is often better practice to use the function disp since this leads to clearer scripts. The disp function allows the display of text and values on the screen. To output the contents of the matrix A on the screen we write disp(A);. Text output must be placed in single quotes, for example:

```
disp('this will display this text');
```

```
[r1,r2]=rootquad(4.2, 6.1, 4);
```
or
```
[p,q]=rootquad(x+2*y,x-y,2*z);
```

or using the function feval

```
[r1,r2]=feval('rootquad',1,2,3);
```

A more important application of feval, which is widely used in this text, is in the process of defining functions which themselves have functions as parameters. These functions can be evaluated internally in the body of the calling function by using feval.

1.17 SOME PITFALLS IN MATLAB

We now list five important points which if observed will enable the MATLAB user to avoid some significant difficulties. This list is not exhaustive.

- It is important to take care when naming files and functions. File and function names follow the rules for variable names, i.e. they must start with a letter followed by a combination of letters or digits and names of existing functions must not be used.

- Do not use MATLAB function names or commands for variable names.

- Matrix sizes are set by assignment so it is vital to ensure that matrix sizes are compatible. Often it is a good idea initially to assign a matrix to an appropriate sized zero matrix; this also makes execution more efficient. For example, consider the following simple script:

```
for i=1:2
  b(i)=i*i;
end
a=[4 5; 6 7];
a*b'
```

  We assign two elements to b in the for loop and make a a 2 x 2 array so that we would expect this script to succeed. However, if b had in the same session been previously set to be a different size matrix then this script will fail. To ensure that it works correctly we must either assign b to be a null matrix using b = [ ], or make b a column vector of two elements by using b = zeros(2,1) or by using the clear statement to clear all variables from the system.

- To create a matrix of zeros that is the same size as a given matrix P the recommended form in MATLAB version 4 is B=zeros(size(P)). To create a P x P matrix, where P is a scalar, the user writes B=zeros(P). Earlier versions of MATLAB allowed the user to write an ambiguous statement of the form B=zeros(P) where P could be a scalar or a matrix.

• Take care with dot products. For example, when defining functions where any of the variables may be vectors then dot products must be used. Note also that 2.^x and 2. ^x are different because the space is important. The first example gives the dot power whilst the second gives 2.0 to the power x, not the dot power. Similar care with spaces must be taken when using complex numbers. For example, a=[1 2-i*4] assigns two elements: 1 and the complex number 2 − i4. In contrast b=[1 2 -i*4] assigns three elements: 1, 2 and the imaginary number −i4.

1.18 SPEEDING UP CALCULATIONS IN MATLAB

Calculations can be greatly speeded up by using vector operations rather than using a loop to repeat a calculation. To illustrate this consider the following simple examples:

Example 1. This script fills the vector **b** using a for loop.

```
% Fill b with square roots of 1 to 1000 using a for loop
clear;
tic;
for i =1:1000
  b(i)=sqrt(i);
end
t=toc;
disp(['Time taken for loop method is ', num2str(t)]);
```

Example 2. This script fills the vector **b** using a vector operation.

```
% Fill b with square roots of 1 to 1000 using a vector
clear;
tic;
a=1:1:1000;
b=sqrt(a);
t=toc;
disp(['Time taken for vector method is ',num2str(t)]);
```

The times taken (in seconds) using various computers were as follows:

	by loop	by vector	ratio
Macintosh LC II	10.92	0.333	33
Macintosh Quadra 610	1.82	0.083	22
PC 486-66MHz	0.74	0.006	123
Sun Sparc LX	0.81	0.005	162

These results illustrate the need to think very carefully about the way algorithms are implemented in MATLAB, particularly with regard to the use of vectors and arrays.

PROBLEMS

1.1.   (a) Start up MATLAB. In the **command** window type x=-1:0.1:1 and then execute each of the following statements by typing them in and pressing return:

```
sqrt(x)              cos(x)
sin(x)               x.^2
x.^3                 plot(x, sin(x.^3))
plot(x, cos(x.^4))
```

Examine the effects of each statement carefully.

(b) Execute the following and explain the results:

```
x=[2 3 4 5]
y=-1:1:2
x.^y
x.*y
x./y
```

1.2.   (a) Set up the matrix A=[1 5 8;84 81 7;12 34 71] in the **command** window and examine the contents of A(1,1), A(2,1), A(1,2), A(3,3), A(1:2,:), A(:,1), A(3,:), A(:,2:3).

(b) What do the following MATLAB statements produce?

```
x=1:1:10
z=rand(10)
y=[z;x]
c=rand(4)
e=[c  eye(size(c)); eye(size(c)) ones(size(c))]
d=sqrt(c)
t1=d*d
t2=d.*d
```

1.3.   Set up a 4 x 4 matrix. Given that the function sum(x) will give the sum of the elements of the vector x, use the function sum to find the sums of the first row and second column of the matrix.

1.4.   Solve the following system of equations using the MATLAB function inv and also using the operators \ and / in the **command** window:

$$\begin{aligned}
2x + y + 5z &= 5 \\
2x + 2y + 3z &= 7 \\
x + 3y + 3z &= 6
\end{aligned}$$

Verify the solution is correct using matrix multiplication.

1.5.   Write a simple script to input two square matrices **A** and **B**: then add, subtract and multiply them. Comment the script and use `disp` to output suitable titles.

1.6.   Write a MATLAB script to set up a 4 x 4 random matrix A and a four-element column vector b. Calculate `x=A\b` and display the result. Calculate `A*x` and compare it with b.

1.7.   Write a simple script to plot the two functions $y_1 = x^2\cos x$ and $y_2 = x^2\sin x$ on the same graph. Use comments in your script and take $x = -2{:}0.1{:}2$.

1.8.   Write a MATLAB script to produce graphs of the functions $y = \cos x$ and $y = \cos(x^3)$ in the range $x = -4{:}0.02{:}4$ using the same axes. Use the MATLAB functions `xlabel`, `ylabel` and `title` to annotate your graphs clearly.

1.9.   Draw the function $y = \exp(-x^2)\cos(20x)$ in the range $x = -2{:}0.1{:}2$. All axes should be labelled and a title included. Compare the results of using the functions `fplot` and `plot` to plot this function.

1.10. Write a MATLAB script to draw the functions $y = 3\sin(\pi x)$ and $y = \exp(-0.2x)$ on the same graph for $x = 0{:}0.02{:}4$. All axes should be labelled. Use `gtext` to label one of the several point of intersection of the graphs.

1.11. Use the functions `mesh` and `meshgrid` to obtain a three-dimensional plot of the function

$$z = 2xy/(x^2 + y^2) \text{ for } x = 1{:}0.1{:}3 \text{ and } y = 1{:}0.1{:}3$$

Redraw the surface using the function `surf` and `contour`.

1.12. An iterative equation for solving the equation $x^2 - x - 1 = 0$ is given by

$$x_{r+1} = 1 + (1/x_r) \text{ for } r = 0, 1, 2, \ldots$$

Given x_0 is 2 write a MATLAB script to solve the equation. Sufficient accuracy is obtained when $\text{abs}(x_{r+1} - x_r) < 0.0005$. Include a check on the answer.

1.13. Given a 4 x 5 matrix **A**, write a script to find the sums of each of the columns using

1. The `for` statement.
2. The function `sum`.

1.14. Given a vector **x** with n elements, write a MATLAB script to form the products

$$p_k = x_1 x_2 \ldots x_{k-1} x_{k+1} \ldots x_n$$

for $k = 1, 2, \ldots, n$. That is, p_k will contain the products of all the vector elements except the kth. Run your script with specific values of x and n.

1.15. The series for $\log_e(1 + x)$ is given by

$$\log_e(1 + x) = x - x^2/2 + x^3/3 - \ldots + (-1)^{k+1} x^k/k \ldots$$

Write a MATLAB script to input a value for x and sum the series while the value of the current term is greater than or equal to the variable *tol*. Use values of $x = 0.5$ and 0.82 and *tol* $= 0.005$ and 0.0005. The result should be checked by using the MATLAB function log. The script should display the value of x and *tol* and the value of $\log_e(1 + x)$ obtained. Use input and disp functions to obtain clear output and prompts.

1.16. Write a MATLAB script to generate a matrix which has the values d along the main diagonal and the values c on the diagonals above and below the main diagonal and zero elsewhere. Your script should allow the user to input any values for c and d and work for any size of matrix n. The script should give clear prompts for input and display the results with a suitable heading.

1.17. Write a MATLAB function to solve the quadratic equation

$$ax^2 + bx + c = 0$$

The function will use three input parameters a, b, c and output the values of the two roots. You should take account of the three cases:

(1) no real roots
(2) real and different roots
(3) equal roots.

Hint: Develop the function rootquad given on page 22.

1.18. Adjust the function of problem 1.17 to deal with the case when $a = 0$. That is, when the equation is non-quadratic. In this case include a third output parameter which will have the value 1 if the equation is quadratic and 0 otherwise.

1.19. Write a simple function to define $f(x) = x^2 - \cos x - x$ and plot the graph of the function in the range 0 to 2. Use this graph to find an initial approximation to the root and then apply the function fzero to find the root to tolerance 0.0005.

2

Linear equations and eigensystems

2.1 INTRODUCTION

When physical systems are modelled mathematically they are sometimes described by linear equation systems or eigensystems and in this chapter we will examine how such equation systems are solved. Linear equation systems can be expressed in terms of matrices and vectors and we introduce some of the more important properties of vectors and matrices in Appendix 1.

MATLAB is an ideal environment for studying linear equation systems because MATLAB functions and operators can work directly on vectors and matrices. It is rich in functions and operators which facilitate the manipulation of matrices. MATLAB originated in an easily accessible implementation of the LINPACK (Dongarra *et al.*, 1979) and EISPACK (Smith *et al.*, 1976, Garbow *et al.*, 1977) routines. These routines were developed specifically to solve linear equations and eigenvalue problems respectively.

We now begin with a discussion of linear equation systems and defer discussion of eigensystems until section 2.15. To illustrate how linear equation systems arise in the modelling of certain physical problems we will consider how current flows are calculated in a simple electrical network. The necessary equations can be developed using one of several techniques; we use the loop-current method together with Ohm's law and Kirchhoff's voltage law. A *loop current* is assumed to circulate around each loop in the network. Thus, in the network given in Fig. 2.1.1, the loop current I_1 circulates around the closed loop *abcd*. Note that the current $I_1 - I_2$ is assumed to flow in the link connecting *b* to *c*. Ohm's law states that the voltage across an ideal resistor

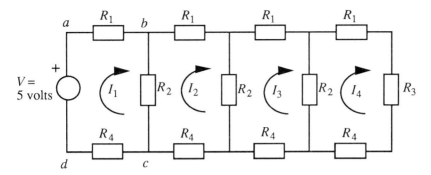

Fig. 2.1.1. Electrical network.

is proportional to the current flow through the resistor. For example, for the link connecting b to c

$$V_{bc} = R_2(I_1 - I_2)$$

where R_2 is the value of the resistor in the link connecting b to c. Kirchhoff's voltage law states that the algebraic sum of the voltages around a loop is zero. Applying these laws to the circuit $abcd$ of Fig. 2.1.1 we have

$$V_{ab} + V_{bc} + V_{cd} = V$$

Substituting the product of current and resistance for voltage gives

$$R_1 I_1 + R_2(I_1 - I_2) + R_4 I_1 = V$$

We can repeat this process for each loop to obtain the following four equations:

$$
\begin{aligned}
(R_1 + R_2 + R_4)I_1 - R_2 I_2 &= V \\
(R_1 + 2R_2 + R_4)I_2 - R_2 I_1 - R_2 I_3 &= 0 \\
(R_1 + 2R_2 + R_4)I_3 - R_2 I_2 - R_2 I_4 &= 0 \\
(R_1 + R_2 + R_3 + R_4)I_4 - R_2 I_3 &= 0
\end{aligned}
\qquad (2.1.1)
$$

Letting $R_1 = R_4 = 1\ \Omega$, $R_2 = 2\ \Omega$, $R_3 = 4\ \Omega$, and $V = 5$ volts, (2.1.1) becomes

$$
\begin{aligned}
4I_1 - 2I_2 &= 5 \\
-2I_1 + 6I_2 - 2I_3 &= 0 \\
-2I_2 + 6I_3 - 2I_4 &= 0 \\
-2I_3 + 8I_4 &= 0
\end{aligned}
$$

This is a system of linear equations in four variables, $I_1, ..., I_4$. In matrix notation it becomes

$$\begin{bmatrix} 4 & -2 & 0 & 0 \\ -2 & 6 & -2 & 0 \\ 0 & -2 & 6 & -2 \\ 0 & 0 & -2 & 8 \end{bmatrix} \begin{bmatrix} I_1 \\ I_2 \\ I_3 \\ I_4 \end{bmatrix} = \begin{bmatrix} 5 \\ 0 \\ 0 \\ 0 \end{bmatrix} \qquad (2.1.2)$$

This equation has the form $\mathbf{Ax} = \mathbf{b}$ where $\mathbf{A}$ is a square matrix of known coefficients, in this case relating to the values of the resistors in the circuit. The vector $\mathbf{b}$ is a vector of known coefficients, in this case the voltage applied to each current loop. The vector $\mathbf{x}$ is the vector of unknown currents. Although this set of equations can be solved by hand, the process is time consuming and error prone. Using MATLAB we simply enter matrix A and vector b and use the command A\b as follows:

```
»A=[4 -2 0 0;-2 6 -2 0;0 -2 6 -2;0 0 -2 8];
»b=[5;0;0;0];
»A\b
ans =
    1.5426
    0.5851
    0.2128
    0.0532
```

The sequence of operations that are invoked by this apparently simple command will be examined in section 2.3.

In many electrical networks the ideal resistors of Fig. 2.1.1 are more accurately represented by electrical impedances. When a harmonic alternating current (AC) supply is connected to the network, electrical engineers represent the impedances by complex quantities. This is to account for the effect of capacitance and/or inductance. To illustrate this we will replace the 5 volt DC supply to the network of Fig. 2.1.1 by a 5 volt AC supply and replace the ideal resistors $R_1, ..., R_4$ by impedances $Z_1, ..., Z_4$. Thus (2.1.1) becomes

$$\begin{aligned} (Z_1 + Z_2 + Z_4)I_1 - Z_2I_2 &= V \\ (Z_1 + 2Z_2 + Z_4)I_2 - Z_2I_1 - Z_2I_3 &= 0 \\ (Z_1 + 2Z_2 + Z_4)I_3 - Z_2I_2 - Z_2I_4 &= 0 \\ (Z_1 + Z_2 + Z_3 + Z_4)I_4 - Z_2I_3 &= 0 \end{aligned} \qquad (2.1.3)$$

At the frequency of the 5 volt AC supply we will assume that $Z_1 = Z_4 = (1 + 0.5j)$ Ω, $Z_2 = (2 + 0.5j)$ Ω, $Z_3 = (4 + 1j)$ Ω where $j = \sqrt{(-1)}$. Electrical engineers prefer to use j rather than i for $\sqrt{(-1)}$. This avoids any possible confusion with I or i which are normally used to denote the current in a circuit. Thus (2.1.3) becomes

$$\begin{aligned} (4 + 1.5j)I_1 - (2 + 0.5j)I_2 &= 5 \\ -(2 + 0.5j)I_1 + (6 + 2j)I_2 - (2 + 0.5j)I_3 &= 0 \\ -(2 + 0.5j)I_2 + (6 + 2j)I_3 - (2 + 0.5j)I_4 &= 0 \\ -(2 + 0.5j)I_3 + (8 + 2.5j)I_4 &= 0 \end{aligned}$$

This system of linear equations becomes, in matrix notation,

$$\begin{bmatrix} (4+1.5j) & -(2+0.5j) & 0 & 0 \\ -(2+0.5j) & (6+2j) & -(2+0.5j) & 0 \\ 0 & -(2+0.5j) & (6+2j) & -(2+0.5j) \\ 0 & 0 & -(2+0.5j) & (8+2.5j) \end{bmatrix} \begin{bmatrix} I_1 \\ I_2 \\ I_3 \\ I_4 \end{bmatrix} = \begin{bmatrix} 5 \\ 0 \\ 0 \\ 0 \end{bmatrix} \qquad (2.1.4)$$

Note that the coefficient matrix is now complex. This does not present any difficulty for MATLAB because the operator A\b works directly with both real and complex numbers. Thus:

```
»p=4+1.5*i;   q=-2-0.5*i;
»r=6+2*i;   s=8+2.5*i;
»A=[p q 0 0;q r q 0;0 q r q;0 0 q s];
»b=[5;0;0;0];
»A\b
ans =
   1.3008 - 0.5560i
   0.4560 - 0.2504i
   0.1530 - 0.1026i
   0.0361 - 0.0274i
```

Note that strictly we have no need to re-enter the values in vector b, assuming that we have not cleared the memory, reassigned the vector b or quit MATLAB. The answer shows that currents flowing in the network are complex. This means that there is a phase difference between the applied harmonic voltage and the currents flowing.

We will now begin a more detailed examination of linear equation systems.

2.2 LINEAR EQUATION SYSTEMS

In general, a linear equation system can be written in matrix form as

$$\mathbf{Ax} = \mathbf{b} \qquad (2.2.1)$$

where $\mathbf{A}$ is an $n \times n$ matrix of known coefficients, $\mathbf{b}$ is a column vector of n known coefficients and $\mathbf{x}$ is the column vector of n unknowns. We have already seen an example of this type of equation system in section 2.1 where the matrix equation (2.1.2) is the matrix equivalent of the linear equations (2.1.1).

The equation system (2.2.1) is called homogeneous if $\mathbf{b} = \mathbf{0}$ and inhomogeneous if $\mathbf{b} \neq \mathbf{0}$. Before attempting to solve an equation system it is reasonable to ask if it has a solution and if so is it unique? A linear inhomogeneous equation system may be *consistent* and have one or an infinity of solutions or be *inconsistent* and have no solution. This is illustrated in Fig. 2.2.1 for a system of three equations in three variables x_1, x_2 and x_3. Each equation represents a plane surface in the x_1, x_2, x_3 space. In Fig. 2.2.1(a) the three planes have a common point of intersection. The coordinates of the point of intersection give the unique solution for the three equations. In Fig. 2.2.1(b) the three planes intersect in a line. Any point on the line of intersection represents a solution so there is no unique solution but an infinite number of solutions satisfying the three equations. Finally in Fig. 2.2.1(c) two of the surfaces are parallel to each other and therefore they never intersect whilst in Fig. 2.2.1(d) each pair of surfaces intersect in different lines. In both of these cases there is no solution and the equations these surfaces represent are inconsistent.

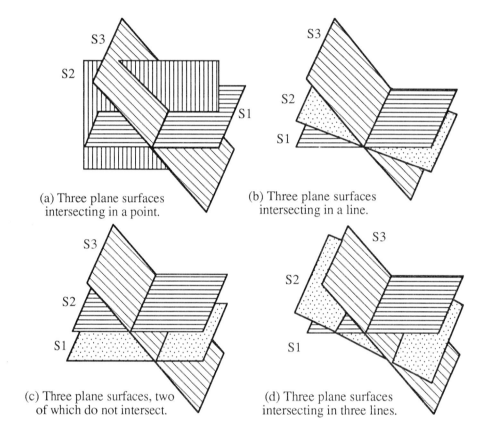

(a) Three plane surfaces intersecting in a point.

(b) Three plane surfaces intersecting in a line.

(c) Three plane surfaces, two of which do not intersect.

(d) Three plane surfaces intersecting in three lines.

Fig. 2.2.1. Three intersecting plane surfaces.

To obtain an algebraic solution to the inhomogeneous equation system (2.2.1) we multiply both sides of (2.2.1) by a matrix called the inverse of $\mathbf{A}$, denoted by $\mathbf{A}^{-1}$, thus:

$$\mathbf{A}^{-1}\mathbf{A}\mathbf{x} = \mathbf{A}^{-1}\mathbf{b}$$

where $\mathbf{A}^{-1}$ is defined by $\mathbf{A}^{-1}\mathbf{A} = \mathbf{A}\,\mathbf{A}^{-1} = \mathbf{I}$. Thus we obtain

$$\mathbf{x} = \mathbf{A}^{-1}\mathbf{b} \tag{2.2.2}$$

The standard algebraic formula for the inverse of **A** is

$$\mathbf{A}^{-1} = \text{adj}(\mathbf{A})/|\mathbf{A}| \qquad\qquad (2.2.3)$$

where $|\mathbf{A}|$ is the determinant of **A** and adj(**A**) is the adjoint of **A**. The determinant and the adjoint of a matrix are defined in Appendix 1. Equations (2.2.2) and (2.2.3) are algebraic statements allowing us to determine **x** but they do not provide an efficient means of solving the system because computing $\mathbf{A}^{-1}$ using (2.2.3) is extremely inefficient, involving order $(n + 1)!$ multiplications where n is the number of equations. However, (2.2.3) is theoretically important because it shows that if $|\mathbf{A}| = 0$ then **A** does not have an inverse. The matrix **A** is then said to be singular and a unique solution for **x** does not exist. Thus establishing that $|\mathbf{A}|$ is non-zero is one way of showing that an inhomogeneous equation system is a consistent system with a unique solution. It is shown in sections 2.6 and 2.7 that (2.2.1) can be solved without formally determining the inverse of **A**.

An important concept in linear algebra is the rank of a matrix. For a square matrix, the rank is the number of independent rows or columns in the matrix. Independence can be explained as follows. The rows (or columns) of a matrix can clearly be viewed as a set of vectors. A set of vectors is said to be linearly independent if none of them can be expressed as a linear combination of any of the others. By linear combination we mean a sum of scalar multiples of the vectors. For example, the matrix

$$\begin{bmatrix} 1 & 2 & 3 \\ -2 & 1 & 4 \\ -1 & 3 & 7 \end{bmatrix} \text{ or } \begin{bmatrix} [\,1 \ 2 \ 3\,] \\ [-2 \ 1 \ 4\,] \\ [-1 \ 3 \ 7\,] \end{bmatrix} \text{ or } \left[\begin{bmatrix} 1 \\ -2 \\ -1 \end{bmatrix} \begin{bmatrix} 2 \\ 1 \\ 3 \end{bmatrix} \begin{bmatrix} 3 \\ 4 \\ 7 \end{bmatrix} \right]$$

has linearly dependent rows and columns because row 3 – row 1 – row 2 = 0 and column 3 – 2(column 2) + column 1 = 0. There is only one equation relating the rows (or columns) and thus there are two independent rows (or columns). Hence this matrix has a rank of 2. Now consider

$$\begin{bmatrix} 1 & 2 & 3 \\ 2 & 4 & 6 \\ 3 & 6 & 9 \end{bmatrix}$$

Here row 2 = 2(row 1) and row 3 = 3(row 1). There are two equations relating the rows and hence only one row is independent so that the matrix has a rank of 1. Note that the number of independent rows and columns in a square matrix is identical, i.e. its row rank and column rank are equal. In general matrices may be non-square and the rank of an $m \times n$ matrix **A** is written rank(**A**). Matrix **A** is said to be of full rank if rank(**A**) = min(m, n); otherwise rank(**A**) < min(m, n) and **A** is said to be rank deficient. MATLAB provides the function rank which works with both square and non-square matrices. In practice, MATLAB determines the rank of a matrix from its singular values, see section 2.10.

For example, consider the following MATLAB statements:

```
»B=[1 2 3;3 4 7;4 -3 1;-2 5 3;1 -7 6]
B =
        1       2       3
        3       4       7
        4      -3       1
       -2       5       3
        1      -7       6

»rank(B)
ans =
        3
```

Thus **B** is of full rank since the rank equals the minimum size of the matrix.

A useful operation in linear algebra is the conversion of a matrix to its reduced row echelon form (RREF). The RREF is defined in Appendix 1. In MATLAB we can use the rref function to compute the RREF of a matrix thus:

```
»rref(B)
ans =
        1       0       0
        0       1       0
        0       0       1
        0       0       0
        0       0       0
```

It is a property of the RREF of a matrix that the number of non-zero rows equals the rank of the matrix. In this example we see that there are only three non-zero rows in the RREF of the matrix confirming that its rank is 3. The RREF also allows us to determine whether a system has a unique solution or not.

We have discussed a number of important concepts relating to the nature of linear equations and their solutions. We now summarise the equivalencies between these concepts. Let **A** be an $n \times n$ matrix. If $\mathbf{Ax} = \mathbf{b}$ is consistent and has a unique solution then:

$\mathbf{Ax} = \mathbf{0}$ has only the trivial solution $\mathbf{x} = \mathbf{0}$.
A is non-singular and $\det(\mathbf{A}) \neq 0$.
The RREF of **A** is the identity matrix.
A has n linearly independent rows and columns.
A has full rank, i.e. $\text{rank}(\mathbf{A}) = n$.

In contrast, if $\mathbf{Ax} = \mathbf{b}$ is either inconsistent or consistent but with more than one solution then:

$\mathbf{Ax} = \mathbf{0}$ has more than one solution.

$\mathbf{A}$ is singular and $\det(\mathbf{A}) = 0$.

The RREF of $\mathbf{A}$ contains at least one zero row.

$\mathbf{A}$ has linearly dependent rows and columns.

$\mathbf{A}$ is rank deficient, i.e. $\text{rank}(\mathbf{A}) < n$.

So far we have only considered the case where there are as many equations as unknowns. Now we consider the cases where there are fewer or more equations than unknown variables.

If there are fewer equations than unknowns then the system is said to be under-determined. The equation system does not have a unique solution; it is either consistent with an infinity of solutions, or inconsistent with no solution. These conditions can be illustrated by reference to Fig. 2.2.1. The diagrams illustrate the nature of the three equations that correspond to the three plane surfaces in the x_1, x_2, x_3 space. If we have a system of only two equations we must imagine the effect of removing one surface from the diagram. Suppose surface S3 is removed. In Fig. 2.2.1(a), (b) and (d) we can see that surfaces S1 and S2 now intersect in a line so that the equations are consistent with an infinity of solutions determined by the line of intersection. In Fig. 2.2.1(c) we see that S1 and S2 do not intersect so that the equations remain inconsistent.

Consider the following system of equations:

$$\begin{bmatrix} 1 & 2 & 3 & 4 \\ -4 & 2 & -3 & 7 \end{bmatrix} \begin{bmatrix} x_1 \\ x_2 \\ x_3 \\ x_4 \end{bmatrix} = \begin{bmatrix} 1 \\ 3 \end{bmatrix}$$

This under-determined system can be rearranged thus:

$$\begin{bmatrix} 1 & 2 \\ -4 & 2 \end{bmatrix} \begin{bmatrix} x_1 \\ x_2 \end{bmatrix} + \begin{bmatrix} 3 & 4 \\ -3 & 7 \end{bmatrix} \begin{bmatrix} x_3 \\ x_4 \end{bmatrix} = \begin{bmatrix} 1 \\ 3 \end{bmatrix}$$

or

$$\begin{bmatrix} 1 & 2 \\ -4 & 2 \end{bmatrix} \begin{bmatrix} x_1 \\ x_2 \end{bmatrix} = \left(\begin{bmatrix} 1 \\ 3 \end{bmatrix} - \begin{bmatrix} 3 & 4 \\ -3 & 7 \end{bmatrix} \begin{bmatrix} x_3 \\ x_4 \end{bmatrix} \right)$$

Thus we have reduced this to a system of two equations in two unknowns, provided values are assumed for x_3 and x_4. Thus the problem has an infinity of solutions, depending on the values chosen for x_3 and x_4.

If a system has more equations than unknowns it is said to be over-determined. We illustrate the situations that may arise by imagining the effect of adding an extra plane surface to the three existing surfaces shown in Fig. 2.2.1. In Fig. 2.2.1(a) the extra surface may pass through the existing point of intersection so that the system remains consistent with a unique solution. Alternatively, the extra surface may miss the existing point of intersection and the system of equations will be inconsistent. In Fig. 2.2.1(b) the extra surface might

(1) intersect the other surfaces in the existing line of intersection so that the system remains consistent with an infinity of solutions,

(2) intersect the existing line of intersection making the system consistent with a unique solution,

(3) miss the existing line of intersection completely, making the system inconsistent.

If an extra surface is added to Fig. 2.2.1(c) or (d) the system will remain inconsistent.

Over-determined systems are often inconsistent, but only marginally so. The inconsistency often arises because the coefficients in the equations are determined experimentally; if the coefficients were known exactly it is likely that the equations would be consistent with a unique solution. Rather than accepting that the system is inconsistent we may ask what the best solution is that satisfies the equations approximately. In the following section we will show how MATLAB deals with this and other categories of linear systems.

2.3 MATLAB OPERATORS \ AND / FOR SOLVING Ax = b

The purpose of this section is to introduce the reader to the MATLAB operator \, including the enhancements implemented in MATLAB v4. A detailed discussion of the algorithms behind its operation will be given in later sections. This operator is a very powerful one which provides a unified approach to the solution of many categories of linear equation systems. The operators / and \ perform matrix "division" and have identical effects. Thus to solve $\mathbf{Ax} = \mathbf{b}$ we may write either A\b or b'/A'. In the latter case the solution $\mathbf{x}$ is expressed as a row rather than a column vector. The operator / or \, when solving $\mathbf{Ax} = \mathbf{b}$, selects the appropriate algorithm dependent on the form of the matrix $\mathbf{A}$. These cases are outlined below:

<u>if</u> $\mathbf{A}$ is a triangular matrix the system is solved by back or forward substitution alone, described in section 2.6.

<u>elseif</u> $\mathbf{A}$ is a positive definite, square symmetric or Hermitian matrix, Cholesky decomposition (described in section 2.8) is applied. When $\mathbf{A}$ is sparse, Cholesky decomposition is preceded by a symmetric minimum degree preordering (described in section 2.14).

<u>elseif</u> $\mathbf{A}$ is a square matrix, general LU decomposition (described in section 2.7) is applied. If $\mathbf{A}$ is sparse this is preceded by a non-symmetric minimum degree preordering (described in section 2.14).

<u>elseif</u> $\mathbf{A}$ is a full non-square matrix, QR decomposition (described in section 2.9) is applied.

<u>elseif</u> $\mathbf{A}$ is a sparse non-square matrix it is augmented and then a minimum degree preordering is applied, followed by sparse Gaussian elimination (described in section 2.14).

The MATLAB \ operator can also be used to solve **AX** = **B** where **B** and the unknown **X** are $m \times n$ matrices. This could provide a simple method of finding the inverse of **A**. If we make **B** the identity matrix **I** then we have

$$\mathbf{AX} = \mathbf{I}$$

and **X** must be the inverse of **A** since $\mathbf{AA}^{-1} = \mathbf{I}$. Thus in MATLAB we could determine the inverse of **A** by using the statement A\eye(size(A)). However, MATLAB provides the function inv(A) to find the inverse of a matrix. It is important to stress that the inverse of a matrix should only be determined if it is specifically required. If we require the solution of a set of linear equations it is more efficient to use the operator \.

We now practically examine some cases to show how the \ operator works, beginning with the solution of a system where the system matrix is triangular. The experiment in this case examines the time and flops taken by the operator \ to solve a system which is full and then when the same system is converted to triangular form by zeroing appropriate elements to produce a triangular matrix. The script used for this experiment is

```
disp(' n    fultime      fflop    fflop/n^3  tritime  triflop
triflop/n^2'');
   a=[ ];
   for n=10:10:50
     a=100*rand(n); b=[1:n]';
     tic; flops(0)
     x=a\b;
     t1=toc; f1=flops;
     f1d=f1/n^3;
     for i=1:n
       for j=i+1:n
         a(i,j)=0;
       end
     end
   tic; flops(0)
   x=a\b;
   t2=toc; f2=flops;
   f2d=f2/n^2;
   fprintf('%2.0f%9.2f%13.2f%9.2f%9.2f%11.2f%9.2f\n',n,t1,f1,f1d,t2,f2,f2d)
   end;
```

The results for a series of randomly generated $n \times n$ matrices are as follows:

n	fultime	fflop	fflop/n^3	tritime	triflop	triflop/n^2
10	0.08	1302.00	1.30	0.05	138.00	1.38
20	0.23	7944.00	0.99	0.05	458.00	1.15
30	0.48	23750.00	0.88	0.08	978.00	1.09
40	0.78	52804.00	0.83	0.12	1698.00	1.06
50	1.27	99122.00	0.79	0.15	2618.00	1.05

Column 1 of this table gives the size of the square matrix, n. To demonstrate that the operator \ takes account of the triangular form, columns 2, 3 and 4 contain execution time, flops and flops divided by n^3 for the full matrix problem. Columns 5, 6 and 7 give the execution time, flops and flops divided by n^2 for the triangular system. These interesting results show that for the full matrix the number of flops taken by \ is proportional to n^3 while for the triangular system the number of flops taken by \ is proportional to n^2. This is the expected result for simple back substitution. In addition we see a considerable saving in time when the operator \ is used with a triangular system.

We now perform experiments to examine the effects of using the operator \ with positive definite symmetric systems. This is a more complex problem than those previously discussed and the script below implements this test. It is based on comparing the application of the \ operator to a positive definite system and a non-positive definite system of equations. For a positive definite system we use the Pascal matrix. To generate a non-positive definite system we add a random matrix to the Pascal matrix. We compare the time and flops taken to solve the two forms of matrix. There is also an important addition which compares the condition numbers of the two matrices. This allows us to check if the condition number of the matrix which is not positive definite is worse or better than the Pascal matrix. The script takes the form

```
disp(' n  post posfl posfl/n^3 rtime rflops rflop/n^3 cr/1000');
for n=10:2:20
  a=[ ];
  a=pascal(n); cd1=cond(a); b=[1:n]';
  tic; flops(0)
  x=a\b;
  t1=toc; f1=flops;
  f1d=f1/n^3;
  a=a+rand(size(a)); cd2=cond(a); cr=cd1/cd2/1000;
  tic; flops(0);
  x=a\b;
  t2=toc; f2=flops;
  f2d=f2/n^3;
fprintf('%2.0f%7.2f%9.2f%7.2f%9.2f%9.2f%7.2f%11.1f\n',n,t1,f1,f1d,t2,f2,f2d,cr)
end;
```

The result of running this script is

n	post	posfl	posfl/n^3	rtime	rflops	rflop/n^3	cr/1000
10	0.05	803.00	0.80	0.07	1326.00	1.33	21.1
12	0.07	1244.00	0.72	0.10	2190.00	1.27	70.5
14	0.07	1817.00	0.66	0.12	3110.00	1.13	1378.7
16	0.08	2538.00	0.62	0.13	4416.00	1.08	23375.1
18	0.15	3423.00	0.59	0.17	6034.00	1.03	168468.8
20	0.12	4488.00	0.56	0.22	7954.00	0.99	3260460.0

Column 1 of this table gives n, the size of the matrix. Columns 2 to 4 give the time, flops and flops divided by n^3 for the positive definite matrix and columns 5 to 7 give the time, flops and flops divided by n^3 for the non-positive definite matrix. The last column gives the ratio of the condition number of the positive definite matrix to the non-positive definite matrix divided by 1000. These results show that although the condition number of the positive definite matrix is much worse than the non-positive definite matrix, the performance of the \ operator for this matrix is significantly better in terms of flops and clearly better in terms of time. This is because the operator \ checks to see if the matrix is positive definite and if so uses the more efficient Cholesky decomposition.

The next test we perform is a rigorous one to examine how the operator \ succeeds with the very badly conditioned Hilbert matrix. The condition of a matrix is a measure of how the accuracy of solutions of linear systems are affected by errors and it is discussed in detail in section 2.4. The test gives the time taken to solve the system, the number of floating point operations required and the accuracy of the solution given by the norm of the residuals, i.e. norm($\mathbf{Ax} - \mathbf{b}$). In addition the test compares these results for the \ operator with the results obtained using the inverse, i.e. $\mathbf{x} = \mathbf{A}^{-1}\mathbf{b}$. The script for this test is

```
disp(' n op time  op flops   op acc   invtime  invflops   invacc');
for n=4:2:14
  a=hilb(n); b=[1:n]';
  tic; flops(0)
  x=a\b;
  t1=toc; f1=flops;
  nm1=norm(b-a*x);
  tic; flops(0)
  x=inv(a)*b;
  t2=toc; f2=flops;
  nm2=norm(b-a*x);
fprintf('%2.0f%8.2f%10.2f%12.2e%8.2f%10.2f%12.2e\n',n,t1,f1,nm1,t2,f2,nm2)
end;
```

This produces the following table of results:

n	op time	op flops	op acc	invtime	invflops	invacc
4	0.03	112.00	4.16e-14	0.15	268.00	1.50e-14
6	0.05	253.00	4.19e-12	0.05	699.00	1.27e-10
8	0.07	478.00	1.91e-10	0.08	1486.00	1.02e-06
10	0.07	803.00	1.16e-08	0.10	2713.00	4.33e-05
12	0.07	1244.00	3.24e-07	0.32	4464.00	5.56e-01
14	0.33	4084.00	2.93e-06	0.37	6843.00	1.47e+01

This output has been edited to remove warnings about the ill-conditioning of the matrix. Column 1 gives the size of the matrix. Columns 2, 3 and 4 give the execution time, flops and accuracy when using the \ operator. Columns 5, 6 and 7 give the same information when using the inv function.

The results in the above table demonstrate convincingly the superiority of the \ operator over the inv function for solving a system of linear equations. It is faster and more accurate than using matrix inversion. However, it should be noted that the accuracy falls off as the matrix becomes increasingly ill-conditioned. In addition we

note that the number of flops required is roughly proportional to n^3. The following
MATLAB statements show this. The variable fpn is filled with the number of flops for
the operator \ taken from the above table and divided by the size of the problem cubed.

```
»fpn=[112 253 478 803 1244 4084]
»fpn./[4 6 8 10 12 14].^3
ans =
    1.7500    1.1713    0.9336    0.8030    0.7199    1.4883
```

These are not constant owing to the effect of overheads when n is small and ill-
conditioning when n is large.

We now use the MATLAB operator \ to solve an under-determined system. Consider
the system

$$x_1 + 2x_2 + 3x_3 = 1$$
$$-4x_1 + 2x_2 - 3x_3 = 3$$

We use MATLAB as follows:

```
»a=[1 2 3;-4 2 -3];
»b=[1;3];
»a\b
ans =
         0
    1.0000
   -0.3333
```

Note that in this case MATLAB has arbitrarily assigned x_1 to zero. MATLAB does not
warn the user that this is only one of an infinity of solutions.

We conclude this section by using the operator \ to solve the following over-
determined system:

$$x_1 + x_2 = 1.98$$
$$2.05x_1 - x_2 = 0.95$$
$$3.06x_1 + x_2 = 3.98$$
$$-1.02x_1 + 2x_2 = 0.92$$
$$4.08x_1 - x_2 = 2.90$$

This is an inconsistent system of equations. To solve this system we use the following
MATLAB script:

```
a=[1 1;2.05 -1;3.06 1;-1.02 2;4.08 -1];
b=[1.98;0.95;3.98;0.92;2.90];
x=a\b
r=(a*x-b)'
```

Running this script gives

```
x =
    0.9631
    0.9885

r =
   -0.0284    0.0358   -0.0444    0.0747    0.0409
```

Any solution to an over-determined system can only approximately satisfy all the equations, as illustrated by the fact that the residuals, calculated from $\mathbf{Ax} - \mathbf{b}$, are non-zero. The operator \ uses a least squares approximation, discussed in detail in section 2.12.

2.4 ACCURACY OF SOLUTIONS AND ILL-CONDITIONING

We now consider factors which affect the accuracy of the solution of $\mathbf{Ax} = \mathbf{b}$ and how any inaccuracies can be detected. We begin with the following examples:

Example 1. Consider the following MATLAB statements:

```
»a=[3.021 2.714 6.913;1.031 -4.273 1.121;5.084 -5.832 9.155]
a =
    3.0210    2.7140    6.9130
    1.0310   -4.2730    1.1210
    5.0840   -5.8320    9.1550

»b=[12.648;-2.121;8.407]
b =
   12.6480
   -2.1210
    8.4070

»a\b
ans =
    1.0000
    1.0000
    1.0000
```

This result is correct and easily verified by substitution into the original equations.

Example 2. Consider Example 1 with $a(2,2)$ changed from -4.2730 to -4.2750 thus:

```
»a(2,2)=-4.2750
a =
    3.0210    2.7140    6.9130
    1.0310   -4.2750    1.1210
    5.0840   -5.8320    9.1550
```

```
»a\b
ans =
   -1.7403
    0.6851
    2.3212
```

Here we have a solution which is very different from that of Example 1, even though the only change in the equation system is less than 0.1% in coefficient $a(2,2)$.

The two examples shown above have dramatically different solutions because they are ill-conditioned. Ill-conditioning can be interpreted graphically by representing each of the equation systems by three plane surfaces, in the manner shown in Fig. 2.2.1. In an ill-conditioned system at least two of the surfaces will be almost parallel so that the point of intersection of the surfaces will be very sensitive to small changes in slope, caused by small changes in coefficient values.

A system of equations is said to be ill-conditioned if a relatively small change in the elements of the coefficient matrix **A** causes a relatively large change in the solution. Conversely a system of equations is said to be well-conditioned if a relatively small change in the elements of the coefficient matrix **A** causes a relatively small change in the solution. Clearly we require a measure of the condition of a system of equations. We know that a system of equations without a solution – the very worst condition possible – has a coefficient matrix with a determinant of zero. It is therefore tempting to think that the size of the determinant of **A** can be used as a measure of condition. However, if **Ax** = **b** and **A** is an $n \times n$ diagonal matrix with each element on the leading diagonal equal to s, then **A** is perfectly conditioned, irrespective of the value of s. But the determinant of **A** in this case is s^n. Thus, the size of the determinant of **A** is not a suitable measure of condition because in this example it changes with s even though the condition of the system is constant.

Two of the functions MATLAB provides to estimate the condition of a matrix are cond and rcond. The function cond is a sophisticated function and is based on singular value decomposition, discussed in section 2.10. For a perfect condition cond is unity but gives a large value for a matrix which is ill-conditioned. The function rcond is less reliable but usually faster. This function gives a value between zero and one. The smaller the value, the worse the conditioning. The reciprocal of rcond is usually of the same order of magnitude as cond. We now illustrate these points with two examples.

Example 3.

```
a =
   20    0    0
    0   20    0
    0    0   20

»[det(a) rcond(a) cond(a)]

ans =
        8000           1           1
```

Example 4.

```
»b=[1 2 3;4 5 6;7 8 9.000001];
»format short e
»[det(b) rcond(b) 1/rcond(b) cond(b)]

ans =
  -3.0000e-06    9.2593e-09    1.0800e+08    1.0109e+08
```

Examples 3 and 4 illustrate perfectly conditioned and badly conditioned matrices respectively. We note that the reciprocal of the rcond value is close to the value of cond. Using the MATLAB functions cond and rcond we now investigate the condition number of the Hilbert matrix, using the script shown below:

```
% Hilbert matrix test.
disp('    n          cond           rcond      log10(cond)')
for n= 4:15
  a=hilb(n);
  fprintf('%5.0f%16.4e',n,cond(a));
  fprintf('%16.4e%10.2f\n',rcond(a),log10(cond(a)));
end
```

Running this script gives

»	n	cond	rcond	log10(cond)
	4	1.5514e+04	4.6812e-05	4.19
	5	4.7661e+05	1.4444e-06	5.68
	6	1.4951e+07	4.4211e-08	7.17
	7	4.7537e+08	1.3478e-09	8.68
	8	1.5258e+10	4.0918e-11	10.18
	9	4.9315e+11	1.2375e-12	11.69
	10	1.6025e+13	3.7331e-14	13.20
	11	5.2161e+14	1.1268e-15	14.72
	12	1.6671e+16	3.5722e-17	16.22
	13	1.7432e+18	1.8052e-18	18.24
	14	3.1905e+17	7.2519e-19	17.50
	15	3.7510e+17	1.2062e-18	17.57

This shows that the Hilbert matrix is ill-conditioned for relatively small values of n, the size of the matrix. The last column of the above output gives the value of $\log_{10}$ of the condition number of the appropriate Hilbert matrix. This gives a rule of thumb estimate of the number of significant figures lost in solving an equation system with this matrix or inverting the matrix.

The next section begins the detailed examination of the algorithms used by the \ operator.

2.5 ELEMENTARY ROW OPERATIONS

We now examine the operations that can usefully be carried out on each equation of a system of equations. Such a system will have the form

$$a_{11}x_1 + a_{12}x_2 + \ldots + a_{1n}x_n = b_1$$

$$a_{21}x_1 + a_{22}x_2 + \ldots + a_{2n}x_n = b_2$$

$$\vdots \qquad \vdots \qquad \qquad \vdots \qquad \vdots$$

$$a_{n1}x_1 + a_{n2}x_2 + \ldots + a_{nn}x_n = b_n$$

or in matrix notation

$$\mathbf{Ax = b}$$

where

$$\mathbf{A} = \begin{bmatrix} a_{11} & a_{12} & \cdots & a_{1n} \\ a_{21} & a_{22} & \cdots & a_{2n} \\ \vdots & \vdots & & \vdots \\ a_{n1} & a_{n2} & \cdots & a_{nn} \end{bmatrix} \qquad \mathbf{b} = \begin{bmatrix} b_1 \\ b_2 \\ \vdots \\ b_n \end{bmatrix} \qquad \mathbf{x} = \begin{bmatrix} x_1 \\ x_2 \\ \vdots \\ x_n \end{bmatrix}$$

A is called the coefficient matrix. Any operation performed on an equation must be applied to both its left- and right-hand side. With this in mind it is helpful to combine the coefficient matrix **A** with the right-hand side vector **b** thus:

$$\begin{bmatrix} a_{11} & a_{12} & \cdots & a_{1n} & b_1 \\ a_{21} & a_{22} & \cdots & a_{2n} & b_2 \\ \vdots & \vdots & & \vdots & \vdots \\ a_{n1} & a_{n2} & \cdots & a_{nn} & b_n \end{bmatrix}$$

This new matrix is called the augmented matrix and we will write it as [**A b**]. We have chosen to adopt this notation because it is consistent with MATLAB notation for combining **A** and **b**. Note that if **A** is an $n \times n$ matrix then the augmented matrix is an $n \times (n + 1)$ matrix. Each row of the augmented matrix holds all the coefficients of an equation and any operation must be applied to every element in the row. The three elementary row operations described below can be applied to individual equations in a system without altering the solution of the equation system. They are:

(1) Interchange the position of any two rows (i.e. equations).

(2) Multiply a row (i.e. equation) by a non-zero scalar.

(3) Replace a row by the sum of the row and a scalar multiple of another row.

These elementary row operations can be used to solve some important problems in linear algebra and we now discuss an application of them.

2.6 SOLUTION OF Ax = b BY GAUSSIAN ELIMINATION

Gaussian elimination is an efficient way to solve equation systems, particularly those with a non-symmetric coefficient matrix having a relatively small number of zero elements. The method depends entirely on using the three elementary row operations, described in section 2.5. Essentially the procedure is to form the augmented matrix for the system and then reduce the coefficient matrix part to an upper triangular form. To illustrate the systematic use of the elementary row operations we consider the application of Gaussian elimination to solve the following equation system:

$$
\begin{bmatrix} 3 & 6 & 9 \\ 2 & (4+p) & 2 \\ -3 & -4 & -11 \end{bmatrix} \begin{bmatrix} x_1 \\ x_2 \\ x_3 \end{bmatrix} = \begin{bmatrix} 3 \\ 4 \\ -5 \end{bmatrix}
\qquad (2.6.1)
$$

where the value of p is known. Table 2.6.1 shows the sequence of operations, beginning at stage 1 with the augmented matrix. In stage 1 the element in the first column of the first row (enclosed in a box in the table) is designated the pivot. We wish to make the elements of column 1 in rows 2 and 3 zero. To achieve this, we divide row 1 by the pivot and then add or subtract a suitable multiple of the modified row 1 to or from rows 2 and 3. The result of this is shown in stage 2 of the table. We then select the next pivot. This is the element in the second column of the new second row, which in the second stage is equal to p. If p is large this does not present a problem but if p is small then numerical problems may arise because we will be dividing all the elements of the new row 2 by this small quantity p. If p is zero then we have an impossible situation because we cannot divide by zero. This difficulty is not related to ill-conditioning; indeed this particular equation system is quite well-conditioned when p is zero. To circumvent

Table 2.6.1. Gaussian elimination to transform
an augmented matrix to upper triangular form.

row 1	$\boxed{3}$	6	9	3	**Stage 1**: Initial matrix.
row 2	2	$(4+p)$	2	4	
row 3	-3	-4	-11	-5	
oldrow 1	3	6	9	3	**Stage 2**: Reduce col 1
oldrow 2 − 2(oldrow 1)/3	0	p	-4	2	of row 2 & 3 to zero.
oldrow 3 + 3(oldrow 1)/3	0	2	-2	-2	
oldrow 1	3	6	9	3	Interchange
oldrow 3	0	$\boxed{2}$	-2	-2	rows 2 & 3.
oldrow 2	0	p	-4	2	
oldrow 1	3	6	9	3	**Stage 3**: Reduce col 2
oldrow 2	0	2	-2	-2	of row 3 to zero.
oldrow 3 − p(oldrow 2)/2	0	0	$\boxed{-4+p}$	$(2+p)$	

these problems the usual procedure is to interchange the row in question with the row containing the element of largest modulus in the column *below the pivot*. In this way we provide a new and larger pivot. This procedure is called partial pivoting. If we assume in this case that p is small or zero then we must interchange rows 2 and 3 as shown in stage 3 of the table to replace p by 2 as the pivot. From row 3 we now subtract row 2 divided by the pivot and multiplied by a coefficient in order to make the element of column 2, row 3, zero. In stage 4 of the table it can be seen that the original coefficient matrix has been reduced to an upper triangular matrix. If $p = 0$ we obtain

$$3x_1 + 6x_2 + 9x_3 = 3 \tag{2.6.2}$$
$$2x_2 - 2x_3 = -2 \tag{2.6.3}$$
$$-4x_3 = 2 \tag{2.6.4}$$

We can now obtain the values of the unknowns x_1, x_2 and x_3 by a process called back substitution. We solve the equations in reversed order. Thus from (2.6.4), $x_3 = -0.5$. From (2.6.3), knowing x_3, we have $x_2 = -1.5$. Finally from (2.6.2), knowing x_2 and x_3, we have $x_1 = 5.5$.

It can be shown that the determinant of a matrix can be evaluated from the product of the elements on the main diagonal provided at stage 3 in Table 2.6.1. This product must be multiplied by $(-1)^m$ where m is the number of row interchanges used. For example, in the above problem, with $p = 0$, one row interchange is used so that $m = 1$ and the determinant of the coefficient matrix is given by $3 \times 2 \times (-4) \times (-1)^1 = 24$.

A method for solving a linear equation system that is closely related to Gaussian elimination is Gauss–Jordan elimination. The method uses the same elementary row operations but differs from Gaussian elimination because elements both below and above the leading diagonal are reduced to zero. This means that back substitution is avoided. For example, solving system (2.6.1) with $p = 0$ leads to the following augmented matrix:

$$\begin{bmatrix} 3 & 0 & 0 & 16.5 \\ 0 & 2 & 0 & -3.0 \\ 0 & 0 & -4 & 2.0 \end{bmatrix}$$

Thus $x_1 = 16.5/3 = 5.5$, $x_2 = -3/2 = -1.5$ and $x_3 = 2/-4 = -0.5$.

Gaussian elimination requires order $n^3/3$ multiplications followed by back substitution requiring order n^2 multiplications. Gauss–Jordan elimination requires order $n^3/2$ multiplications. Thus Gauss–Jordan elimination requires approximately 50% more operations than Gaussian elimination.

2.7 LU DECOMPOSITION

LU decomposition (or factorisation) is a similar process to Gaussian elimination and is equivalent in terms of elementary row operations. The matrix A can be decomposed so that

$$A = LU \qquad\qquad\qquad\qquad (2.7.1)$$

where L is a lower triangular matrix with a leading diagonal of ones and U is an upper triangular matrix. Matrix A may be real or complex. Compared with Gaussian elimination, LU decomposition has a particular advantage when the equation system we wish to solve, $Ax = b$, has more than one right-hand side or when the right-hand sides are not known in advance. This is because the factors L and U are obtained explicitly and they can be used for any right-hand sides as they arise without recalculating L and U. Gaussian elimination does not determine L explicitly but rather forms $L^{-1}b$ so that all right-hand sides must be known when the equation is solved.

The major steps required to solve an equation system by LU decomposition are as follows. Since $A = LU$ then $Ax = b$ becomes

$$LUx = b$$

where b is not restricted to a single column. Putting $y = Ux$ leads to

$$Ly = b$$

Because L is a lower triangular matrix this equation is solved efficiently by forward substitution. To find x we now solve

$$Ux = y$$

Because U is an upper triangular matrix this equation can also be solved efficiently by back substitution.

We now illustrate the LU decomposition process by solving (2.6.1) with $p = 1$. We are not concerned with b and we do not form an augmented matrix. We proceed exactly as with Gaussian elimination, see Table 2.6.1, except that we keep a record of the elementary row operations performed at the ith stage in $T^{(i)}$ and place the results of these operations in a matrix $U^{(i)}$ rather than over-writing A.

We begin with the matrix

$$A = \begin{bmatrix} 3 & 6 & 9 \\ 2 & 5 & 2 \\ -3 & -4 & -11 \end{bmatrix}$$

Following the same operations as used in Table 2.6.1, we will create a matrix $U^{(1)}$ with zeros below the leading diagonal in the first column using the following elementary row operations:

$$\text{row 2 of } U^{(1)} = \text{row 2 of } A - 2(\text{row 1 of } A)/3 \tag{2.7.2}$$

and

$$\text{row 3 of } U^{(1)} = \text{row 3 of } A + 3(\text{row 1 of } A)/3 \tag{2.7.3}$$

Now A can be expressed as the product $T^{(1)} U^{(1)}$ as follows:

$$\begin{bmatrix} 3 & 6 & 9 \\ 2 & 5 & 2 \\ -3 & -4 & -11 \end{bmatrix} = \begin{bmatrix} 1 & 0 & 0 \\ 2/3 & 1 & 0 \\ -1 & 0 & 1 \end{bmatrix} \begin{bmatrix} 3 & 6 & 9 \\ 0 & 1 & -4 \\ 0 & 2 & -2 \end{bmatrix}$$

Note that row 1 of A and row 1 of $U^{(1)}$ are identical. Thus row 1 of $T^{(1)}$ has a unit entry in column 1 and zero elsewhere. The remaining rows of $T^{(1)}$ are determined from (2.7.2) and (2.7.3). For example, row 2 of $T^{(1)}$ is derived by rearranging (2.7.2) thus:

$$\text{row 2 of } A = \text{row 2 of } U^{(1)} + 2(\text{row 1 of } A)/3$$

or

$$\text{row 2 of } A = 2(\text{row 1 of } U^{(1)})/3 + \text{row 2 of } U^{(1)}$$

since row 1 of $U^{(1)}$ is identical to row 1 of A. Hence row 2 of $T^{(1)}$ is [2/3 1 0].

We now move to the next stage of the decomposition process. In order to bring the largest element of column 2 in $U^{(1)}$ onto the leading diagonal we must interchange rows 2 and 3. Thus $U^{(1)}$ becomes the product $T^{(2)} U^{(2)}$ as follows:

$$\begin{bmatrix} 3 & 6 & 9 \\ 0 & 1 & -4 \\ 0 & 2 & -2 \end{bmatrix} = \begin{bmatrix} 1 & 0 & 0 \\ 0 & 0 & 1 \\ 0 & 1 & 0 \end{bmatrix} \begin{bmatrix} 3 & 6 & 9 \\ 0 & 2 & -2 \\ 0 & 1 & -4 \end{bmatrix}$$

Finally, to complete the process of obtaining an upper triangular matrix we make

$$\text{row 3 of } U = \text{row 3 of } U^{(2)} - (\text{row 2 of } U^{(2)})/2$$

Hence $U^{(2)}$ becomes the product $T^{(3)} U$ as follows:

$$\begin{bmatrix} 3 & 6 & 9 \\ 0 & 2 & -2 \\ 0 & 1 & -4 \end{bmatrix} = \begin{bmatrix} 1 & 0 & 0 \\ 0 & 1 & 0 \\ 0 & 1/2 & 1 \end{bmatrix} \begin{bmatrix} 3 & 6 & 9 \\ 0 & 2 & -2 \\ 0 & 0 & -3 \end{bmatrix}$$

Thus $A = T^{(1)} T^{(2)} T^{(3)} U$ implying that $L = T^{(1)} T^{(2)} T^{(3)}$ as follows:

$$\begin{bmatrix} 1 & 0 & 0 \\ 2/3 & 1 & 0 \\ -1 & 0 & 1 \end{bmatrix} \begin{bmatrix} 1 & 0 & 0 \\ 0 & 0 & 1 \\ 0 & 1 & 0 \end{bmatrix} \begin{bmatrix} 1 & 0 & 0 \\ 0 & 1 & 0 \\ 0 & 1/2 & 1 \end{bmatrix} = \begin{bmatrix} 1 & 0 & 0 \\ 2/3 & 1/2 & 1 \\ -1 & 1 & 0 \end{bmatrix}$$

Note that owing to the row interchanges L is not strictly a lower triangular matrix but it can be made so by interchanging rows.

MATLAB implements LU factorisation by using the function lu and may produce a matrix that is not strictly a lower triangular matrix. However, a permutation matrix **P** may be produced, if required, such that **LU** = **PA** with **L** lower triangular.

We now show how the MATLAB function lu deals with the above example:

```
»a=[3 6 9;2 5 2;-3 -4 -11]
a =
      3       6       9
      2       5       2
     -3      -4     -11
```

To obtain the **L** and **U** matrices we must use that MATLAB facility of assigning two parameters simultaneously as follows:

```
»[L1,U]=lu(a)
L1 =
    1.0000        0        0
    0.6667   0.5000   1.0000
   -1.0000   1.0000        0

U =
    3.0000   6.0000   9.0000
         0   2.0000  -2.0000
         0        0  -3.0000
```

Note that the L1 matrix is not in lower triangular form, although its true form can easily be deduced by interchanging rows to form a triangle. To obtain a true lower triangular matrix we must assign three parameters as follows:

```
»[L,U,P]=lu(a)
L =
    1.0000        0        0
   -1.0000   1.0000        0
    0.6667   0.5000   1.0000

U =
    3.0000   6.0000   9.0000
         0   2.0000  -2.0000
         0        0  -3.0000

P =
      1       0       0
      0       0       1
      0       1       0
```

In the above output P is the permutation matrix such that LU=PA or P'LU=A. Thus PL is equal to L1.

The MATLAB operator \ determines the solution of **Ax** = **b** using LU factorisation. As an example of an equation system with multiple right-hand sides we solve **AX** = **B** where

$$A = \begin{bmatrix} 3 & 4 & -5 \\ 6 & -3 & 4 \\ 8 & 9 & -2 \end{bmatrix} \text{ and } B = \begin{bmatrix} 1 & 3 \\ 9 & 5 \\ 9 & 4 \end{bmatrix}$$

Performing LU decomposition, such that **LU = A**, gives

$$L = \begin{bmatrix} 0.375 & -0.064 & 1 \\ 0.750 & 1 & 0 \\ 1 & 0 & 0 \end{bmatrix} \text{ and } U = \begin{bmatrix} 8 & 9 & -2 \\ 0 & -9.75 & 5.5 \\ 0 & 0 & -3.897 \end{bmatrix}$$

Thus **LY = B** is

$$\begin{bmatrix} 0.375 & -0.064 & 1 \\ 0.750 & 1 & 0 \\ 1 & 0 & 0 \end{bmatrix} \begin{bmatrix} y_{11} & y_{12} \\ y_{21} & y_{22} \\ y_{31} & y_{32} \end{bmatrix} = \begin{bmatrix} 1 & 3 \\ 9 & 5 \\ 9 & 4 \end{bmatrix}$$

We note that implicitly we have two systems of equations which when separated can be written

$$\left[A\right] \begin{bmatrix} y_{11} \\ y_{21} \\ y_{31} \end{bmatrix} = \begin{bmatrix} 1 \\ 9 \\ 9 \end{bmatrix} \text{ and } \left[A\right] \begin{bmatrix} y_{12} \\ y_{22} \\ y_{32} \end{bmatrix} = \begin{bmatrix} 3 \\ 5 \\ 4 \end{bmatrix}$$

In this example **L** is not strictly a lower triangular matrix owing to the reordering of the rows. However, the solution of this equation is still found by forward substitution. For example, $1y_{11} = b_{31} = 9$, so that $y_{11} = 9$. Then $0.75y_{11} + 1y_{21} = b_{21} = 9$. Hence $y_{21} = 2.25$. The complete **Y** matrix is

$$Y = \begin{bmatrix} 9.000 & 4.000 \\ 2.250 & 2.000 \\ -2.231 & 1.628 \end{bmatrix}$$

Finally solving **UX = Y** by back substitution gives

$$X = \begin{bmatrix} 1.165 & 0.891 \\ 0.092 & -0.441 \\ 0.572 & -0.418 \end{bmatrix}$$

The MATLAB function **det** determines the determinant of a matrix using LU factorisation as follows. Since $A = LU$ then $|A| = |L| |U|$. The elements of the leading diagonal of **L** are all ones so that $|L| = 1$. Since **U** is upper triangular, its determinant is the product of the elements of its leading diagonal. Thus taking account of row interchanges the appropriately signed product of the diagonal elements of **U** gives the determinant.

2.8 CHOLESKY DECOMPOSITION

Cholesky decomposition or factorisation is a form of triangular decomposition that can only be applied to positive definite symmetric or positive definite Hermitian matrices. A symmetric or Hermitian matrix **A** is said to be positive definite if $x^T A x > 0$ for any non-zero **x**. A more useful definition of a positive definite matrix is one that has all eigenvalues greater than zero. The eigenvalue problem is discussed in section 2.15. If **A** is symmetric or Hermitian we can write

$$\mathbf{A} = \mathbf{P}^T\mathbf{P} \quad (\text{or } \mathbf{P}^H\mathbf{P} \text{ when } \mathbf{A} \text{ is Hermitian}) \tag{2.8.1}$$

where **P** is an upper triangular matrix. The algorithm computes **P** row by row by equating coefficients of each side of (2.8.1). Thus p_{11}, p_{12}, p_{13}, ..., p_{22}, p_{23}, ... are determined in sequence, ending with p_{nn}. Coefficients on the leading diagonal of **P** are computed from expressions that involve determining a square root. For example:

$$p_{22} = \sqrt{a_{22} - p_{12}^2}$$

A property of positive definite matrices is that the term under the square root is always positive and so the square root will be real. Furthermore, row interchanges are not required because the dominant coefficients will be on the main diagonal. The whole process requires only about half as many multiplications as LU decomposition. Cholesky factorisation is implemented for positive definite symmetric matrices in MATLAB by the function chol. For example, consider the Cholesky factorisation of the following positive definite Hermitian matrix:

```
»A=[2 -i 0;i 2 0;0 0 3]
A =
    2.0000                0 - 1.0000i              0
    0 + 1.0000i      2.0000                        0
    0                     0                   3.0000

»P=chol(A)
P =
    1.4142                0 - 0.7071i              0
    0                1.2247                        0
    0                     0                   1.7321
```

When the operator \ detects a symmetric positive definite or Hermitian positive definite system matrix it solves $\mathbf{A}\mathbf{x} = \mathbf{b}$ using the following sequence of operations. **A** is factorised into $\mathbf{P}^T\mathbf{P}$, **y** is set to **Px**; then $\mathbf{P}^T\mathbf{y} = \mathbf{b}$. The algorithm solves for **y** by forward substitution since $\mathbf{P}^T$ is a lower triangular matrix. Then **x** can be determined from **y** by backward substitution since **P** is an upper triangular matrix. We can illustrate the steps in this process by the following example:

$$A = \begin{bmatrix} 2 & 3 & 4 \\ 3 & 6 & 7 \\ 4 & 7 & 10 \end{bmatrix} \text{ and } b = \begin{bmatrix} 2 \\ 4 \\ 8 \end{bmatrix}$$

Then by Cholesky factorisation

$$P = \begin{bmatrix} 1.414 & 2.121 & 2.828 \\ 0 & 1.225 & 0.817 \\ 0 & 0 & 1.155 \end{bmatrix}$$

Now since $P^T y = b$ solving for y by forward substitution gives

$$y = \begin{bmatrix} 1.414 \\ 0.817 \\ 2.887 \end{bmatrix}$$

Finally solving $Px = y$ by back substitution gives

$$x = \begin{bmatrix} -2.5 \\ -1.0 \\ 2.5 \end{bmatrix}$$

We now compare the performance of the operator \ with the function chol. Clearly their performance should be similar in the case of a positive definite matrix. The Pascal matrix, created by the function pascal, is both symmetric and positive definite. The function pascal is used in the following script:

```
    disp('n   op t    op fl  op fl/n^3   choltime  cholflops  cholflop/
n^3');
    for n=10:2:20
      a=[ ];
      a=pascal(n);
      b=[1:n]';
      tic;flops(0)
      x=a\b;
      t1=toc;f1=flops;
      f1d=f1/n^3;
      tic;flops(0);
      r=chol(a);v=r'\b;
      x=r\b;
      t2=toc;f2=flops;
      f2d=f2/n^3;
     fprintf('%2.0f%6.2f%10.2f%10.2f%10.2f%12.2f%10.2f\n',n,t1,f1,f1d,
t2,f2,f2d)
    end;
```

Running this script gives

n	op t	op fl	op fl/n^3	choltime	cholflops	cholflop/n^3
10	0.05	803.00	0.80	0.07	643.00	0.64
12	0.05	1244.00	0.72	0.08	1004.00	0.58
14	0.08	1817.00	0.66	0.10	1481.00	0.54
16	0.08	2538.00	0.62	0.12	2090.00	0.51
18	0.10	3423.00	0.59	0.15	2847.00	0.49
20	0.12	4488.00	0.56	0.17	3768.00	0.47

The similarity in performance of the function chol and the operator \ is borne out by the above table. In this table, column 1 is the size of the matrix and columns 2, 3 and 4 give the execution time, flops and flops/n^3 using the \ operator. Columns 5, 6 and 7 give the same information using Cholesky decomposition explicitly to solve the same problem.

Cholesky factorisation *can* be applied to a symmetric matrix which is not positive definite but the process does not possess the numerical stability of the positive definite case. Furthermore one or more rows in **P** may be purely imaginary. For example:

$$\text{If } \mathbf{A} = \begin{bmatrix} 1 & 2 & 3 \\ 2 & -5 & 9 \\ 3 & 9 & 6 \end{bmatrix} \text{ then } \mathbf{P} = \begin{bmatrix} 1 & 2 & 3 \\ 0 & 3i & -i \\ 0 & 0 & 2i \end{bmatrix}$$

This is not implemented in MATLAB.

2.9 QR DECOMPOSITION

We have seen how a square matrix can be decomposed or factorised into the product of a lower and an upper triangular matrix by the use of elementary row operations. An alternative decomposition is to an upper triangular matrix and an orthogonal matrix if **A** is real or a unitary matrix if **A** is complex. This is called QR decomposition. Thus

$$\mathbf{A} = \mathbf{QR}$$

where **R** is the upper triangular matrix and **Q** is the orthogonal or unitary matrix. If **Q** is orthogonal $\mathbf{Q}^{-1} = \mathbf{Q}^T$ and if **Q** is unitary $\mathbf{Q}^{-1} = \mathbf{Q}^H$, very useful properties.

There are several procedures which provide QR decomposition; here we present Householder's method. To decompose a real matrix, Householder's method begins by defining a matrix **P** thus:

$$\mathbf{P} = \mathbf{I} - 2\mathbf{w}\mathbf{w}^T \tag{2.9.1}$$

Providing $\mathbf{w}^T\mathbf{w} = 1$ then **P** is both symmetrical and orthogonal. The orthogonality can be easily verified by expanding the product $\mathbf{P}^T\mathbf{P}$ (= **PP**) as follows:

$$\mathbf{PP} = (\mathbf{I} - 2\mathbf{w}\mathbf{w}^T)(\mathbf{I} - 2\mathbf{w}\mathbf{w}^T)$$

$$= \mathbf{I} - 4\mathbf{w}\mathbf{w}^T + 4\mathbf{w}\mathbf{w}^T(\mathbf{w}\mathbf{w}^T) = \mathbf{I}$$

To decompose $\mathbf{A}$ into $\mathbf{QR}$, we begin by forming the vector $\mathbf{w}_1$ from the coefficients of the first column of $\mathbf{A}$ as follows:

$$\mathbf{w}_1^T = \mu_1[(a_{11} - s_1) \ a_{21} \ a_{31} \ \ldots \ a_{n1}]$$

where

$$\mu_1 = \frac{1}{\sqrt{2s_1 (s_1 - a_{11})}} \quad \text{and} \quad s_1 = \pm\left(\sum_{j=1}^{n} a_{j1}^2\right)^{1/2}$$

By substituting for μ_1 and s_1 in $\mathbf{w}_1$ it can be verified that the necessary orthogonality condition, $\mathbf{w}_1^T\mathbf{w}_1 = 1$, is satisfied. Substituting $\mathbf{w}_1$ into (2.9.1) we generate an orthogonal matrix $\mathbf{P}^{(1)}$.

The matrix $\mathbf{A}^{(1)}$ is now created from the product $\mathbf{P}^{(1)}\mathbf{A}$. It can easily be verified that all elements in the first column of $\mathbf{A}^{(1)}$ are zero except for the element on the leading diagonal which is equal to s_1. Thus

$$\mathbf{A}^{(1)} = \mathbf{P}^{(1)}\mathbf{A} = \begin{bmatrix} s_1 & + & \cdots & + \\ 0 & + & \cdots & + \\ \vdots & \vdots & & \vdots \\ 0 & + & \cdots & + \\ 0 & + & \cdots & + \end{bmatrix}$$

In the $\mathbf{A}^{(1)}$ matrix, + indicates a non-zero element.

We now begin the second stage of the orthogonalisation process by forming $\mathbf{w}_2$ from the coefficients of the second column of $\mathbf{A}^{(1)}$ thus:

$$\mathbf{w}_2^T = \mu_2[0 \ \ (a_{22}^{(1)} - s_2) \ a_{32}^{(1)} \ a_{42}^{(1)} \ \cdots \ a_{n2}^{(1)}]$$

where a_{ij} are the coefficients of $\mathbf{A}$ and

$$\mu_2 = \frac{1}{\sqrt{2s_2(s_2 - a_{22}^{(1)})}} \quad \text{and} \quad s_2 = \pm\left\{\sum_{j=2}^{n} \left(a_{j2}^{(1)}\right)^2\right\}^{1/2}$$

Then the orthogonal matrix $\mathbf{P}^{(2)}$ is generated from

$$\mathbf{P}^{(2)} = \mathbf{I} - 2\mathbf{w}_2\mathbf{w}_2^T$$

The matrix $\mathbf{A}^{(2)}$ is then created from the product $\mathbf{P}^{(2)}\mathbf{A}^{(1)}$ as follows:

$$\mathbf{A}^{(2)} = \mathbf{P}^{(2)}\mathbf{A}^{(1)} = \mathbf{P}^{(2)}\mathbf{P}^{(1)}\mathbf{A} = \begin{bmatrix} s_1 & + & \cdots & + \\ 0 & s_2 & \cdots & + \\ \vdots & \vdots & & \vdots \\ 0 & 0 & \cdots & + \\ 0 & 0 & \cdots & + \end{bmatrix}$$

Note that $\mathbf{A}^{(2)}$ has zero elements in its first two columns except for the elements on and above the leading diagonal. We can continue this process $n - 1$ times until we obtain an upper triangular matrix $\mathbf{R}$. Thus

$$\mathbf{R} = \mathbf{P}^{(n-1)} ... \mathbf{P}^{(2)}\mathbf{P}^{(1)}\mathbf{A} \qquad\qquad (2.9.2)$$

Note that since $\mathbf{P}^{(i)}$ is orthogonal, the product $\mathbf{P}^{(n-1)} ... \mathbf{P}^{(2)}\mathbf{P}^{(1)}$ is also orthogonal.

We wish to determine the orthogonal matrix $\mathbf{Q}$ such that $\mathbf{A} = \mathbf{QR}$. Thus $\mathbf{R} = \mathbf{Q}^{-1}\mathbf{A}$ or $\mathbf{R} = \mathbf{Q}^{T}\mathbf{A}$. Hence, from (2.9.2),

$$\mathbf{Q}^{T} = \mathbf{P}^{(n-1)} ... \mathbf{P}^{(2)}\mathbf{P}^{(1)}$$

Apart from the signs associated with the columns of $\mathbf{Q}$ and the rows of $\mathbf{R}$ the decomposition is unique. These signs are dependent on whether the positive or negative square root is taken in determining s_1, s_2, etc. Complete decomposition of the matrix requires $2n^3/3$ multiplications and n square roots. To illustrate this procedure consider the decomposition of the matrix

$$\mathbf{A} = \begin{bmatrix} 4 & -2 & 7 \\ 6 & 2 & -3 \\ 3 & 4 & 4 \end{bmatrix}$$

Thus

$$s_1 = \sqrt{(4^2 + 6^2 + 3^2)} = 7.8102$$
$$\mu_1 = 1/\sqrt{\{2 \times 7.8102 \times (7.8102 - 4)\}} = 0.1296$$
$$\mathbf{w}_1^T = 0.1296[(4 - 7.8102)\ \ 6\ \ 3] = [-0.4939\ \ 0.7777\ \ 0.3889]$$

Using (2.9.1) we generate $\mathbf{P}^{(1)}$ and hence $\mathbf{A}^{(1)}$ thus:

$$\mathbf{P}^{(1)} = \begin{bmatrix} 0.5121 & 0.7682 & 0.3841 \\ 0.7682 & -0.2097 & -0.6049 \\ 0.3841 & -0.6049 & 0.6976 \end{bmatrix}$$

$$\mathbf{A}^{(1)} = \mathbf{P}^{(1)}\mathbf{A} = \begin{bmatrix} 7.8102 & 2.0486 & 2.8168 \\ 0 & -4.3753 & 3.5873 \\ 0 & 0.8123 & 7.2936 \end{bmatrix}$$

Note that we have reduced the elements of the first column of $\mathbf{A}^{(1)}$ below the leading diagonal to zero. We continue with the second stage thus:

$$s_2 = \sqrt{\{(-4.3753)^2 + 0.8123^2\}} = 4.4501$$

$$\mu_2 = 1/\sqrt{\{2 \times 4.4501 \times (4.4501 + 4.3753)\}} = 0.1128$$

$$\mathbf{w}_2^T = 0.1128[0\ \ (-4.3753 - 4.4501)\ \ 0.8123] = [0\ \ -0.9958\ \ 0.0917]$$

$$\mathbf{P}^{(2)} = \begin{bmatrix} 1 & 0 & 0 \\ 0 & -0.9832 & 0.1825 \\ 0 & 0.1825 & 0.9832 \end{bmatrix}$$

$$\mathbf{R} = \mathbf{A}^{(2)} = \mathbf{P}^{(2)}\mathbf{A}^{(1)} = \begin{bmatrix} 7.8102 & 2.0486 & 2.8168 \\ 0 & 4.4501 & -2.1956 \\ 0 & 0 & 7.8259 \end{bmatrix}$$

Note that we have now reduced the first two columns of $\mathbf{A}^{(2)}$ below the leading diagonal to zero. This completes the process to determine the upper triangular matrix $\mathbf{R}$. Finally we determine the orthogonal matrix $\mathbf{Q}$ as follows:

$$\mathbf{Q} = (\mathbf{P}^{(2)}\mathbf{P}^{(1)})^{\mathrm{T}} = \begin{bmatrix} 0.5121 & -0.6852 & 0.5179 \\ 0.7682 & 0.0958 & -0.6330 \\ 0.3841 & 0.7220 & 0.5754 \end{bmatrix}$$

It is not necessary for the reader to carry out the above calculations since MATLAB provides the function qr to carry out this decomposition. For example:

```
»a=[4 -2 7;6 2 -3;3 4 4]
a =
     4    -2     7
     6     2    -3
     3     4     4

»[Q,R]=qr(a)
Q =
   -0.5121    0.6852    0.5179
   -0.7682   -0.0958   -0.6330
   -0.3841   -0.7220    0.5754
R =
   -7.8102   -2.0486   -2.8168
         0   -4.4501    2.1956
         0         0    7.8259
```

One advantage of QR decomposition is that it can be applied to non-square matrices, decomposing an $m \times n$ matrix into an $m \times m$ orthogonal matrix and an $m \times n$ upper triangular matrix. Note that if $m > n$ the decomposition is not unique.

2.10 SINGULAR VALUE DECOMPOSITION

The singular value decomposition (SVD) of an $m \times n$ matrix $\mathbf{A}$ is given by

$$\mathbf{A} = \mathbf{U}\mathbf{S}\mathbf{V}^{\mathrm{T}} \text{ (or } \mathbf{U}\mathbf{S}\mathbf{V}^{\mathrm{H}} \text{ if } \mathbf{A} \text{ is complex)}$$

where $\mathbf{U}$ is an orthogonal $m \times m$ matrix and $\mathbf{V}$ is an orthogonal $n \times n$ matrix. If $\mathbf{A}$ is complex then $\mathbf{U}$ and $\mathbf{V}$ are unitary matrices. In all cases $\mathbf{S}$ is a real diagonal $m \times n$ matrix. The elements of the leading diagonal of this matrix are called the singular values of $\mathbf{A}$.

Normally they are arranged in decreasing value so that $s_1 > s_2 > ... > s_n$. Thus

$$\mathbf{S} = \begin{bmatrix} s_1 & 0 & \cdots & 0 \\ 0 & s_2 & \cdots & 0 \\ \vdots & \vdots & & \vdots \\ 0 & 0 & \cdots & s_n \\ 0 & 0 & \cdots & 0 \\ \vdots & \vdots & & \vdots \\ 0 & 0 & \cdots & 0 \end{bmatrix}$$

The singular values are the non-negative square roots of the eigenvalues of $\mathbf{A}^T\mathbf{A}$. Since $\mathbf{A}^T\mathbf{A}$ is symmetric or Hermitian these eigenvalues are real and non-negative so that the singular values are also real and non-negative. Algorithms for computing the SVD of a matrix are given by Golub and Van Loan (1989).

The SVD of a matrix has several important applications. In section 2.4 we introduced the reduced row echelon form of a matrix and explained how the MATLAB function rref gave information from which the rank of a matrix can be deduced. However, rank can be more effectively determined from the SVD of a matrix since its rank is equal to the number of its non-zero singular values. Thus for a 5 x 5 matrix of rank 3, s_4 and s_5 would be zero. In practice, rather than counting the non-zero singular values, MATLAB determines rank from the SVD by counting the number of singular values greater than some tolerance value. This is a more realistic approach to determining rank than counting any non-zero value, however small.

To illustrate how singular value decomposition helps us to examine the properties of a matrix we will use the MATLAB function svd to carry out a singular value decomposition and compare it with the function rref. Consider the following example in which a Vandermonde matrix is created using the MATLAB function vander. The Vandermonde matrix is known to be ill-conditioned. SVD allows us to examine the nature of this ill-conditioning. In particular a zero or a very small singular value would indicate rank deficiency and this example shows that the singular values are becoming relatively close to this condition. In addition SVD allows us to compute the condition number of the matrix. In fact the MATLAB function cond uses SVD to compute the condition number and this gives the same values as obtained by dividing the largest singular value by the smallest singular value. Even the norm of the matrix is supplied by the first singular value. Comparing the SVD with the RREF process in the following script, we see that the result of using the MATLAB functions rref and rank gives the rank of this special Vandermonde matrix as 5 but tells us nothing else. There is no warning that the matrix is badly conditioned.

```
»c=[1 1.01 1.02 1.03 1.04];
»v=vander(c)
v =
      1.0000      1.0000      1.0000      1.0000      1.0000
      1.0406      1.0303      1.0201      1.0100      1.0000
      1.0824      1.0612      1.0404      1.0200      1.0000
      1.1255      1.0927      1.0609      1.0300      1.0000
      1.1699      1.1249      1.0816      1.0400      1.0000

»format long
»s=svd(v)
s =
    5.21036705103790
    0.10191833587669
    0.00069969883945
    0.00000235238029
    0.00000000329498

»norm(v)
ans =
    5.21036705103790

»cond(v)
ans =
     1.581303243779974e+09

»s(1)/s(5)
ans =
     1.581303243779974e+09

»rank(v)
ans =
     5

»rref(v)
ans =
     1      0      0      0      0
     0      1      0      0      0
     0      0      1      0      0
     0      0      0      1      0
     0      0      0      0      1
```

The following example is very similar to the one above but the Vandermonde matrix has now been generated to be rank deficient. The smallest singular value, although not zero, is zero to machine precision and the rank function returns the value of 4.

```
»c=[1 1.01 1.02 1.03 1.03];
»v=vander(c)
v =
      1.0000      1.0000      1.0000      1.0000      1.0000
      1.0406      1.0303      1.0201      1.0100      1.0000
      1.0824      1.0612      1.0404      1.0200      1.0000
      1.1255      1.0927      1.0609      1.0300      1.0000
      1.1255      1.0927      1.0609      1.0300      1.0000

»format long e
»svd(v)
ans =
      5.187797954424028e+00
      8.336322098941451e-02
      3.997349250040803e-04
      8.462129966297493e-07
      1.803237524824342e-23

»format short
»rank(v)
ans =
      4

»rank(v, 1e-24)
ans =
      5

»rref(v)
ans =

      1.0000           0           0           0     -0.9424
           0      1.0000           0           0      3.8262
           0           0      1.0000           0     -5.8251
           0           0           0      1.0000      3.9414
           0           0           0           0           0

»cond(v)
ans =
   2.8769e+23
```

The rank function does allow the user to vary the tolerance. However, tolerance should be used with care since the rank function counts the number of singular values greater than tolerance and this gives the rank of the matrix. If tolerance is very small, i.e. smaller than the machine precision, the rank may be miscounted.

2.11 THE PSEUDO-INVERSE

If A is an $m \times n$ rectangular matrix such that $m > n$, then the system

$$Ax = b \qquad (2.11.1)$$

is an over-determined system of equations. We cannot invert A, since it is not a square matrix, but premultiplying (2.11.1) by A^T converts the system matrix to a square matrix as follows:

$$A^T A x = A^T b$$

Thus

$$x = (A^T A)^{-1} A^T b \qquad (2.11.2)$$

Letting $A^+ = (A^T A)^{-1} A^T$, where A^+ is called the pseudo-inverse of A, the solution of (2.11.1) is given by

$$x = (A^+) b \qquad (2.11.3)$$

This definition requires A to have full rank. A^+ is an $m \times n$ array which is unique. If A is square and non-singular then $A^+ = A^{-1}$. If A is complex then

$$A^+ = (A^H A)^{-1} A^H \qquad (2.11.4)$$

where A^H is the conjugate transpose, described in Appendix 1. The product $A^T A$ has a condition number which is the square of the condition number of A. If A is close to rank deficient then A^+ is best calculated from the SVD of A. If A is real, the SVD of A is USV^T. Thus the SVD of A^T is VS^TU^T so that

$$A^T A = (VS^T U^T)(USV^T) = VS^T SV^T \text{ since } U^T U = I$$

Hence

$$A^+ = (VS^T SV^T)^{-1} VS^T U^T = V^{-T}(S^T S)^{-1} V^{-1} VS^T U^T$$

$$= V(S^T S)^{-1} S^T U^T \qquad (2.11.5)$$

since $V^{-T} = (V^T)^{-1} = (V^T)^T = V$. Since V is an $m \times m$ matrix, U is an $n \times n$ matrix and S is an $n \times m$ matrix then (2.11.5) is conformable, i.e. matrix multiplication is possible, see Appendix 1.

We now consider the case when A is rank deficient. In this case $S^T S$ cannot be inverted because of the very small or zero singular values. To deal with this problem we take only the r non-zero singular values of the matrix so that S is an $r \times r$ matrix where r is the rank of A. To make the multiplications of (2.11.5) conformable we take the first r columns of V and the first r rows of U^T, i.e. the first r columns of U. For example:

```
»a=[1 2 3;4 5 9;7 11 18;-2 3 1;7 1 8]
a =
       1       2       3
       4       5       9
       7      11      18
      -2       3       1
       7       1       8

»rank(a)
ans =
       2

»[u,s,v]=svd(a)
u =
    1.3808e-01    8.3900e-02    9.6760e-01    1.8072e-01   -7.0588e-02
    4.1151e-01    2.1527e-02    9.0709e-02   -5.2001e-01    7.4267e-01
    8.2576e-01    2.7323e-01   -1.7054e-01   -2.5279e-02   -4.6233e-01
    5.2431e-02    5.6503e-01   -1.5013e-01    6.7567e-01    4.4600e-01
    3.5633e-01   -7.7368e-01   -6.2419e-02    4.8966e-01    1.7546e-01

s =
    2.6839e+01             0             0
             0    6.1358e+00             0
             0             0    2.6750e-15
             0             0             0
             0             0             0

v =
    3.7087e-01   -7.2741e-01    5.7735e-01
    4.4452e-01    6.8489e-01    5.7735e-01
    8.1539e-01   -4.2523e-02   -5.7735e-01

»ss=s(1:2,1:2)
ss =
    2.6839e+01             0
             0    6.1358e+00

»v(:,1:2)*inv(ss'*ss)*ss'*u(:,1:2)'
ans =
   -8.0383e-03    3.1342e-03   -2.0981e-02   -6.6261e-02    9.6645e-02
    1.1652e-02    9.2183e-03    4.4174e-02    6.3938e-02   -8.0457e-02
    3.6136e-03    1.2353e-02    2.3193e-02   -2.3230e-03    1.6187e-02
```

It is not necessary to carry out the above operation, since the MATLAB function pinv provides an identical result.

2.12 OVER-DETERMINED SYSTEMS

Consider the following over-determined system of linear equations:

$$
\begin{aligned}
x_1 + x_2 &= 1.98 \\
2.05x_1 - x_2 &= 0.95 \\
3.06x_1 + x_2 &= 3.98 \\
-1.02x_1 + 2x_2 &= 0.92 \\
4.08x_1 - x_2 &= 2.90
\end{aligned}
\tag{2.12.1}
$$

The coefficient matrix of this system is clearly not square. Although over-determined systems may have a unique solution, most often we are concerned with equation systems that are generated from experimental data which can lead to a relatively small degree of inconsistency. Fig. 2.12.1 shows that (2.12.1) is such a system; the lines do not intersect in a point, although there is a point that *nearly* satisfies all the equations.

We would like to choose the best point of all in the region of the intersections. One criterion for doing this is that the point should minimise the sum of squares of the residuals of the equations. The pseudo-inverse, discussed in section 2.11, actually finds this point. However, we can solve the system directly using the operator \ and here we compare the results from these two methods. The MATLAB script below solves system (2.12.1).

```
a=[1 1;2.05 -1;3.06 1;-1.02 2;4.08 -1];
b=[1.98;0.95;3.98;0.92;2.90];
x=pinv(a)*b
norm_pinv=norm(a*x-b)
x=a\b
norm_op=norm(a*x-b)
```

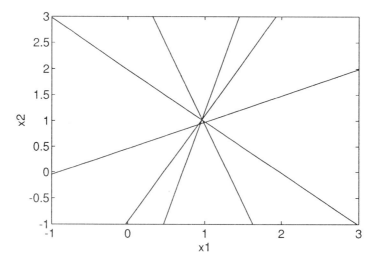

Fig. 2.12.1. Plot of inconsistent equation system (2.12.1).

Running this script gives the following numeric output:

```
x =
    0.9631
    0.9885

norm_pinv =
    0.1064

x =
    0.9631
    0.9885

norm_op =
    0.1064
```

Here both the MATLAB operator \ and the function pinv have provided the same "best fit" solution for the inconsistent set of equations. Fig. 2.12.2 shows the region where these equations intersect in greater detail than Fig. 2.12.1. The symbol '+' indicates the MATLAB solution which can be seen to lie in this region. The MATLAB operator \ does not solve an over-determined system by using the pseudo-inverse, as shown in (2.11.3). Instead, it solves (2.11.1) directly by QR decomposition. QR decomposition can be applied to both square and rectangular matrices providing the number of rows is greater than the number of columns. For example, applying the MATLAB function qr to solve the

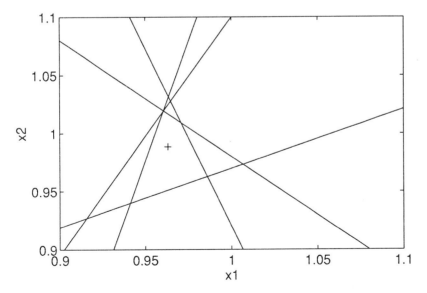

Fig. 2.12.2. Plot of inconsistent equation system (2.12.1)
showing the region of intersection of the equations,
where "+" indicates "best" solution.

over-determined system (2.12.1) we have

```
»a=[1 1;2.05 -1;3.06 1;-1.02 2;4.08 -1];
»b=[1.98;0.95;3.98;0.92;2.90];
»[Q,R]=qr(a)
Q =
    -0.1761     0.4123    -0.7157    -0.2339    -0.4818
    -0.3610    -0.2702     0.0998     0.6751    -0.5753
    -0.5388     0.5083     0.5991    -0.2780    -0.1230
     0.1796     0.6839    -0.0615     0.6363     0.3021
    -0.7184    -0.1756    -0.3394     0.0857     0.5749

R =
    -5.6792     0.7237
          0     2.7343
          0          0
          0          0
          0          0
```

In the equation $\mathbf{Ax} = \mathbf{b}$ we have replaced $\mathbf{A}$ by $\mathbf{QR}$ so that $\mathbf{QRx} = \mathbf{b}$. Let $\mathbf{Rx} = \mathbf{y}$. Then $\mathbf{y} = \mathbf{Q}^{-1}\mathbf{b} = \mathbf{Q}^T\mathbf{b}$ since $\mathbf{Q}$ is orthogonal. Once $\mathbf{y}$ is determined we can efficiently determine $\mathbf{x}$ by back substitution since $\mathbf{R}$ is upper triangular. Thus, continuing the above example,

```
»y=Q'*b
y =
    -4.7542
     2.7029
     0.0212
    -0.0942
    -0.0446
```

Using the second row of $\mathbf{R}$ and the second row of $\mathbf{y}$ we can determine x_2. From the first row of $\mathbf{R}$ and the first row of $\mathbf{y}$ we can determine x_1 since x_2 is known. Thus

$$-5.6792x_1 + 0.7237x_2 = -4.7542$$
$$2.7343x_2 = 2.7029$$

give $x_1 = 0.9631$ and $x_2 = 0.9885$, as before. The MATLAB operator \ implements this sequence of operations.

We conclude this section by considering a case where the coefficient matrix of the over-determined system is rank deficient. The following example is rank deficient and represents a system of parallel lines.

```
»a=[1 2;1 2;1 2;1 2]
a =
     1     2
     1     2
     1     2
     1     2
```

```
»b=[1 1.03 .97 1.01]';
»y=a\b
Warning: Rank deficient, rank = 1  tol =    3.5527e-15

y =
         0
    0.5012

»norm(y)

ans =
    0.5012
```

The user is warned that this system is rank deficient. We have solved the system using the \ operator and now solve it using the pinv function as follows:

```
»x=pinv(a)*b

x =
    0.2005
    0.4010
»norm(x)
ans =
    0.4483
```

We see that when the pinv function and \ operator are applied to rank deficient systems the pinv function gives the solution with the smallest norm.

2.13 ITERATIVE METHODS

Except in special circumstances it is unlikely that any function or script developed by the user will outperform a function or operator that is an integral part of MATLAB. Thus we cannot expect to develop a function that will determine the solution of $\mathbf{Ax} = \mathbf{b}$ more efficiently than by using the MATLAB operation A\b. However, we describe iterative methods here for the sake of completeness.

Iterative methods of solution are developed as follows. We begin with a system of linear equations

$$a_{11}x_1 + a_{12}x_2 + \ldots + a_{1n}x_n = b_1$$
$$a_{21}x_1 + a_{22}x_2 + \ldots + a_{2n}x_n = b_2$$
$$\vdots \qquad \vdots \qquad \qquad \vdots \qquad \vdots$$
$$a_{n1}x_1 + a_{n2}x_2 + \ldots + a_{nn}x_n = b_n$$

These can be rearranged to give

$$x_1 = (b_1 - a_{12}x_2 - a_{13}x_3 - \dots - a_{1n}x_n)/a_{11}$$
$$x_2 = (b_2 - a_{21}x_1 - a_{23}x_3 - \dots - a_{2n}x_n)/a_{22}$$

and so on until

$$x_n = (b_n - a_{n1}x_1 - a_{n2}x_2 - \dots - a_{n-1,n}x_{n-1})/a_{nn}$$

If we assume initial values for x_i, where $i = 1, \dots, n$, and substitute these values into the right-hand side of the above equations we may determine new values for the x_i on the left-hand side of the equations. The iterative process is continued by substituting these values of x_i into the right-hand side of the equations, etc. There are several variants of the process. For example, we can use old values of x_i in the right-hand side of the equations to determine *all* the new values of x_i, in the left-hand side of the equation. This is called Jacobi or simultaneous iteration. Alternatively, we may use a new value of x_i in the right-hand side of the equation as soon as it is determined to obtain the other values of x_i in the right-hand side. This is called Gauss–Seidel or cyclic iteration.

The conditions for convergence for this type of iteration are

$$|a_{ii}| \gg \sum_{\substack{j=1 \\ j \neq i}}^{n} |a_{ij}| \quad \text{for } i = 1, 2, \dots, n$$

Thus these iterative methods are only guaranteed to work when the coefficient matrix is diagonally dominant.

2.14 SPARSE MATRICES

Sparse matrices arise in many problems of science and engineering, for example in linear programming and the analysis of structures. Indeed, most large matrices that arise in the analysis of physical systems are sparse and the recognition of this fact makes the solution of linear systems with millions of coefficients feasible. The aim of this section is to give a brief description of the extensive sparse matrix facilities available in MATLAB v4 and to give practical illustrations of their value through examples. For background information on how MATLAB implements the concept of sparsity, see Gilbert *et al.* (1992). A theoretical justification for the ideas introduced here is beyond the scope of this book.

A matrix is sparse if it contains a high proportion of zero elements. However, this will be practically significant only if the sparsity is of such an extent that we can practically utilise this feature to reduce the computation time and storage facilities required for operations used on such matrices. One major way in which time can be saved in dealing with sparse matrices is to avoid unnecessary operations on zero elements. Thus it is difficult to give a simple quantitative answer to the question: when is a matrix sparse?

MATLAB *does not automatically treat a matrix as sparse* and the sparsity features of MATLAB are not introduced until invoked. Thus the user determines whether a matrix is in the sparse class or the full class. If the user considers a matrix to be sparse and wants to use this fact to advantage the matrix must first be converted to sparse form. This is achieved by using the function sparse. Thus b=sparse(a) converts the matrix a to sparse form and subsequent MATLAB operations will take account of this sparsity. If we wish to return this matrix to full form we simply use c = full(b). However, the sparse function can also be used directly to generate sparse matrices.

It is important to note that binary operators *, +, -, / and \ produce sparse results if *both* operands are sparse. Thus the property of sparsity may survive a long sequence of matrix operations. In addition such functions as chol(A), qr(A) and lu(A) produce sparse results if the matrix A is sparse. However, in mixed cases, where one operand is sparse and the other is full, the result is generally a full matrix. Thus the property of sparsity may be inadvertently lost. Notice in particular that eye(n) is not in the sparse class of matrices in MATLAB but a sparse identity matrix can be created using speye(n). Thus the latter should be used in manipulations with sparse matrices. Fortunately it easy to test if the result of a series of matrix operations is sparse or not by using the issparse function described below and its use is recommended if in doubt.

We will now introduce some of the key MATLAB functions for dealing with sparse matrices, describe their use and, where appropriate, give examples of their application. The simplest MATLAB function which helps in dealing with sparsity is the function nnz(a) which provides the number of non-zero elements in a given matrix a. A function which enables us to examine whether a given matrix has been defined or has been propagated as sparse is the function issparse(a) which returns the value 1 if the matrix a is sparse or 0 if it is not sparse. The function spy(a) allows the user to view the structure of a given matrix a by displaying symbolically only its non-zero elements, see Fig. 2.14.1 below for examples.

Before we can illustrate the action of these and other functions it is useful to generate some sparse matrices. This is easily done by using a different form of the sparse function. This time the function is supplied with the location of the non-zero entries in the matrix, the value of these entries, the size of the sparse matrix and the space allocated for the non-zero entries. This function call takes the form sparse(i,j,nzvals,m,n,nzmax). This allocates the non-zero values in the vector nzvals to the positions in the matrix given by the vectors i and j, the row position being given by i and the column position by j. Space is allocated for nzmax non-zeros. Since all but one parameter is optional there are many forms of this function. We cannot give examples of all these forms but the following cases illustrate its use.

```
»colpos=[1 2 1 2 5 3 4 3 4 5];
»rowpos=[1 1 2 2 2 4 4 5 5 5];
»val=[12 -4 7 3 -8 -13 11 2 7 -4];
»a=sparse(rowpos,colpos,val,5,5)
```

These statements give the following output:

```
a =
   (1,1)        12
   (2,1)         7
   (1,2)        -4
   (2,2)         3
   (4,3)       -13
   (5,3)         2
   (4,4)        11
   (5,4)         7
   (2,5)        -8
   (5,5)        -4
```

We see that a 5 x 5 sparse matrix with 10 non-zero elements has been generated with the required coefficient values in the required positions. This sparse matrix can be converted to a full matrix as follows:

```
»b=full(a)
b =
   12    -4     0     0     0
    7     3     0     0    -8
    0     0     0     0     0
    0     0   -13    11     0
    0     0     2     7    -4
```

This is the equivalent full matrix. Now the following statements test to see if the matrices a and b are in the sparse class and give the number of non-zeros they contain.

```
»[issparse(a) issparse(b) nnz(a) nnz(b)]

ans =
    1     0    10    10
```

Clearly these functions give the expected results. Since a is a member of the class of sparse matrices the value of issparse(a) is 1. However, although b looks sparse it is not *stored* as a sparse matrix and hence is not in the class sparse within the MATLAB environment. The next example shows how to generate a larger 100 x 100 sparse matrix and compares the time taken to solve a linear system of equations involving this sparse matrix with the time taken using the equivalent full matrix. The script for this is

```
%generate a sparse triple diagonal matrix
rowpos=2:100;colpos=1:99;
values=2*ones(1,99);
offdiag=sparse(rowpos,colpos,values,100,100);
a=sparse(1:100,1:100,4*ones(1,100),100,100);
a=a+offdiag+offdiag';
%generate full matrix
b=full(a);
%generate arbitrary right hand side for system of equations
rhs=[1:100]';
%time it                                    [Script continues...
```

```
tic;x=a\rhs;t1=toc;
tic;x=b\rhs;t2=toc;
fprintf('time sparse matrix solve= %4.2f\n',t1);
fprintf('time full matrix solve= %4.2f\n',t2)
```

This provides the results

```
»time sparse matrix solve= 0.68
time full matrix solve= 3.20
```

In this example, using a sparse class of matrix is approximately five times as fast as using a full matrix. We now perform a similar exercise, this time to determine the lu decomposition of a 100 x 100 matrix:

```
offdiag=sparse(2:100,1:99,2*ones(1,99),100,100);
a=sparse(1:100,1:100,4*ones(1,100),100,100);
a=a+offdiag+offdiag';
%generate full matrix
b=full(a);
%generate arbitrary right hand side for system of equations
rhs=[1:100]';
%time
tic; lu1=lu(a); t1=toc;
tic; lu2=lu(b); t2=toc;
fprintf('time sparse lu= %4.2f\n',t1);
fprintf('time full lu= %4.2f\n',t2)
```

The times for this are

```
time sparse lu= 0.40
time full lu= 4.82
```

Again this provides a considerable saving in time.

An alternative way to generate sparse matrices is to use the functions sprandn and sprandsym. These provide random sparse matrices and random sparse symmetric matrices respectively. The call

```
A=sprandn(m,n,d)
```

produces an m x n random matrix with normally distributed non-zero entries of density d. The density is the proportion of the non-zero entries to the total number of entries in the matrix. Thus d must be in the range 0 to 1. To produce a symmetric random matrix with normally distributed non-zero entries of density d we use

```
A=sprandsym(n,d)
```

Examples of calls of these functions are given by

```
»a=sprandn(5,5,0.25)

a =
   (3,1)        0.8717
   (1,3)        0.2641
   (3,3)       -0.7012
   (4,4)       -1.4462
   (1,5)        0.0591
   (2,5)        1.7971

»b=full(a)
b =
        0          0     0.2641          0     0.0591
        0          0          0          0     1.7971
   0.8717          0    -0.7012          0          0
        0          0          0    -1.4462          0
        0          0          0          0          0

»as=sprandsym(5,.25)
as =
   (4,1)        0.5774
   (3,2)       -0.3600
   (2,3)       -0.3600
   (3,3)        1.2460
   (4,3)       -0.6390
   (1,4)        0.5774
   (3,4)       -0.6390

»full(as)
ans =
        0          0          0     0.5774          0
        0          0    -0.3600          0          0
        0    -0.3600     1.2460    -0.6390          0
   0.5774          0    -0.6390          0          0
        0          0          0          0          0
```

An alternative call for sprandsym is given by

    A=sprandsym(n,density,r)

If r is a scalar then this produces a random sparse symmetric matrix with a condition number equal to $1/r$. Remarkably, if r is a vector of length n, a random sparse matrix with eigenvalues equal to the elements of r is produced. A positive definite matrix has all its eigenvalues positive and consequently such a matrix can be generated by choosing each of the n elements of r to be positive. An example of this form of call is

    »aposdef=sprandsym(6,.4,[1 2.5 6 9 2 4.3])

```
aposdef =
    (1,1)        6.3609
    (3,1)       -1.6685
    (4,1)       -0.9371
    (6,1)        2.5369
    (2,2)        2.5000
    (1,3)       -1.6685
    (3,3)        3.4695
    (6,3)        1.8615
    (1,4)       -0.9371
    (4,4)        4.6653
    (6,4)       -0.8399
    (5,5)        2.0000
    (1,6)        2.5369
    (3,6)        1.8615
    (4,6)       -0.8399
    (6,6)        5.8043

»posdeful=full(aposdef)
posdeful =
    6.3609          0    -1.6685    -0.9371         0     2.5369
         0     2.5000          0          0         0          0
   -1.6685          0     3.4695          0         0     1.8615
   -0.9371          0          0     4.6653         0    -0.8399
         0          0          0          0    2.0000          0
    2.5369          0     1.8615    -0.8399         0     5.8043
```

This provides an important method for generating test matrices with required properties since by providing a list of eigenvalues with a range of values we can produce positive definite matrices that are very badly conditioned.

We now return to examine further the value of using sparsity. The reasons for the very high level of improvement in computing efficiency when using the \ operator, illustrated in the example at the beginning of this section, are complex. The process includes a special preordering of the columns of the matrix. This special preordering, called *minimum degree ordering*, is used in the case of the \ operator. This preordering takes different forms depending on whether the matrix is symmetric or non-symmetric. The aim of any preordering is to reduce the amount of *fill-in* from any subsequent matrix operations. Fill-in is the introduction of additional non-zero elements.

We can examine this preordering process using the spy function and the function symmmd which implements *symmetric minimum degree ordering* in MATLAB. The function is automatically applied when working on matrices which belong to the class of sparse matrices for the standard functions and operators of MATLAB. However, if we require to use this preordering in non-standard applications then we may use the symmmd function. The following examples illustrate the use of this function.

We first consider the simple process of multiplication applied to a full and a sparse matrix. The sparse multiplication uses the minimum degree ordering. The following script generates a sparse matrix, obtains a minimum degree ordering for it, and then examines the result of multiplying the matrix by itself transposed. This is compared

with the same operations carried out on the full matrix and the number of flops required
for each operation is compared.

```
%generate a sparse matrix
offdiag=sparse(2:100,1:99,2*ones(1,99),100,100);
offdiag2=sparse(4:100,1:97,3*ones(1,97),100,100);
offdiag3=sparse(95:100,1:6,7*ones(1,6),100,100);
a=sparse(1:100,1:100,4*ones(1,100),100,100);
a=a+offdiag+offdiag'+offdiag2+offdiag2'+offdiag3+offdiag3';
a=a*a';
%generate full matrix
b=full(a);
morder=symmmd(a);
%time & flops
tic; flops(0);
spmult=a(morder,morder)*a(morder,morder)';
t1=toc; flsp=flops;
tic;flops(0);
fulmult=b*b';
t2=toc; flful=flops;
fprintf('time sparse mult= %4.2f flops sparse mult
=%6.0f\n',t1,flsp);
fprintf('time full mult= %4.2f flops full mult= %6.0f\n',t2,flful)
```

Running this script results in the following output:

```
»time sparse mult= 1.78 flops sparse mult=    25562
time full mult= 9.52 flops full mult=   2000018
```

We now perform a similar experiment to that above but for a more complex
numerical process than multiplication. In the script below we examine LU decomposition.
We consider the result of using a minimum degree ordering on the LU decomposition
process by comparing the performance of the lu function with and without a preordering.
The script has the form

```
%generate a sparse matrix
offdiag=sparse(2:100,1:99,2*ones(1,99),100,100);
offdiag2=sparse(4:100,1:97,3*ones(1,97),100,100);
offdiag3=sparse(95:100,1:6,7*ones(1,6),100,100);
a=sparse(1:100,1:100,4*ones(1,100),100,100);
a=a+offdiag+offdiag'+offdiag2+offdiag2'+offdiag3+offdiag3';
a=a*a';a1=flipud(a);a =a+a1;
%generate full matrix
b=full(a);
morder=symmmd(a);
%time & flops
tic; flops(0);
lud=lu(a(morder,morder)); t1=toc;flsp=flops;
tic;flops(0);
fullu=lu(b); t2=toc;flful=flops;
```

[Script continues...

```
subplot(2,2,1), spy(a);title('Original matrix');
subplot(2,2,2), spy(a(morder,morder));title('Ordered Matrix')
subplot(2,2,3), spy(fullu);title('LU decomposition,unordered matrix');
subplot(2,2,4), spy(lud);title('LU decomposition, ordered matrix');
fprintf('time sparse lu= %4.2f flops sparse lu= %6.0f\n',t1,flsp);
fprintf('time full lu= %4.2f flops full lu= %6.0f\n',t2,flful)
```

Running this script gives

```
time sparse lu= 2.45 flops sparse lu=    29647
time full lu= 3.75 flops full lu=    601938
```

As expected we achieve a modest time saving. Fig. 2.14.1 shows the original matrix, the reordered matrix and the LU decomposition structure both with and without minimum degree ordering. Notice that the number of non-zeros in the LU matrices with

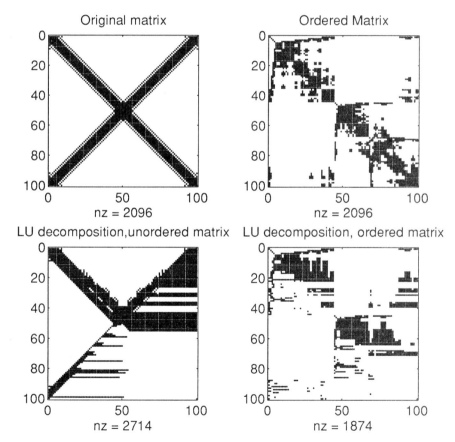

Fig. 2.14.1. Effect of minimum degree ordering on LU decomposition.
The spy function shows the matrix, the ordered matrix
and LU decomposition with and without ordering.

preordering is 1874 and without is 2714. Thus there are over 800 more non-zero elements in the full LU matrices. The number of non-zero elements in the original matrix is 2096. This means that LU decomposition of the full matrix has produced a fill-in of more than 700 extra non-zero elements. In contrast LU decomposition of the preordered matrix has produced less non-zeros than the original matrix. The reduction of fill-in is an important feature of sparse numerical processes and may ultimately lead to great saving in computational effort.

The MATLAB function symmmd provides a minimum degree ordering for symmetric matrices. For non-symmetric matrices MATLAB provides the function colmmd which gives the column minimum degree ordering for non-symmetric matrices. An alternative ordering, which is used to reduce bandwidth, is the reverse Cuthill MacGee ordering.

This is implemented in MATLAB by the function symrcm. The execution of the statement p=symrcm(A) provides the permutation vector p to produce the required preordering and A(p,p) is the reordered matrix.

We have shown that in general taking account of sparsity will provide savings in time. However, these savings fall off as the matrices on which we are operating become less sparse as the following example illustrates.

```
n=100; b=1:n;
disp('   density    timesparse    timefull');
for density=0.01:0.01:0.10
  A=sprandsym(n,density)+0.1*speye(n);
  density=density+1/n;
  tic; x=A\b'; t1= toc;
  B=full(A);
  tic; y=B\b';t2= toc;
  fprintf('%10.3f%12.2f%12.2f\n',density,t1,t2);
end
```

In the above script a diagonal of elements has been added to the randomly generated sparse matrix. This is done to ensure that each row of the matrix contains a non-zero element, otherwise the matrix may be singular. Adding this diagonal modifies the density. If the original $n \times n$ matrix has a density of d, then, assuming that this matrix has only zeros on the diagonal, the modified density is $d + 1/n$.

```
»    density    timesparse    timefull
     0.020        0.80         6.92
     0.030        1.28         6.97
     0.040        1.93         6.77
     0.050        3.62         6.62
     0.060        5.17         6.88
     0.070        6.12         6.72
     0.080        7.32         6.93
     0.090        8.35         6.77
     0.100        8.72         6.75
     0.110        8.50         7.02
```

This output shows that the advantage of using a sparse class of matrix diminishes as the density increases.

Another application where sparsity is important is in solving the least squares problem. This problem is known to be ill-conditioned and hence any saving in computational effort is particularly beneficial. This is directly implemented by using A\b where A is non-square and sparse. To illustrate the use of the \ operator with sparse matrices and compare its performance when no account is taken of sparsity we use the following script:

```
%generate a sparse triple diagonal matrix
rowpos=2:100; colpos=1:99;
values=ones(1,99);
offdiag=sparse(rowpos,colpos,values,100,100);
a=sparse(1:100,1:100,4*ones(1,100),100,100);
a=a+offdiag+offdiag';
%Now generate a sparse least squares system
als=a(:,1:50);
%generate full matrix
cfl=full(als);
rhs=1:100;
tic;x=als\rhs';t1=toc;
tic;x=cfl\rhs';t2=toc;
fprintf('time sparse least squares solve= %4.2f\n',t1);
fprintf('time full least squares solve= %4.2f\n',t2);
```

This provides the results

```
»time sparse least squares solve= 1.83
time full least squares solve= 11.02
```

Again we see the advantage of using sparsity.

We have not covered all aspects of sparsity nor described all the related functions. However, we hope this section provides a helpful introduction to this difficult but important and valuable development of MATLAB.

2.15 THE EIGENVALUE PROBLEM

Eigenvalue problems arise in many branches of science and engineering. For example, the vibration characteristics of structures are determined from the solution of an algebraic eigenvalue problem. Here we consider a particular example of a system of masses and springs shown in Fig. 2.15.1. The equations of motion for this system are

$$m_1\ddot{q}_1 + (k_1 + k_2 + k_4)q_1 - k_2q_2 - k_4q_3 = 0$$

$$m_2\ddot{q}_2 - k_2q_1 + (k_2 + k_3)q_2 - k_3q_3 = 0 \qquad (2.15.1)$$

$$m_3\ddot{q}_3 - k_4q_1 - k_3q_2 + (k_3 + k_4)q_3 = 0$$

where m_1, m_2 and m_3 are the system masses and k_1, ..., k_4 are the spring stiffnesses. If

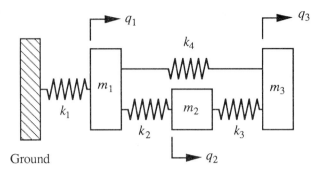

Fig. 2.15.1. Mass–spring system with three degrees of freedom.

we assume an harmonic solution for each coordinate then $q_i(t) = u_i \exp(j\omega t)$ where $j = \sqrt{(-1)}$, for $i = 1, 2$ and 3. Hence $d^2 q_i/dt^2 = -\omega^2 u_i \exp(j\omega t)$. Substituting in (2.15.1) and cancelling the common factor $\exp(j\omega t)$ gives

$$-\omega^2 m_1 u_1 + (k_1 + k_2 + k_4)u_1 - k_2 u_2 - k_4 u_3 = 0$$

$$-\omega^2 m_2 u_2 - k_2 u_1 + (k_2 + k_3)u_2 - k_3 u_3 = 0 \qquad (2.15.2)$$

$$-\omega^2 m_3 u_3 - k_4 u_1 - k_3 u_2 + (k_3 + k_4)u_3 = 0$$

If $m_1 = 10$ kg, $m_2 = 20$ kg, $m_3 = 30$ kg, $k_1 = 10$ kN/m, $k_2 = 20$ kN/m, $k_3 = 25$ kN/m and $k_4 = 15$ kN/m then (2.15.2) becomes

$$-\omega^2 10 u_1 + 45000 u_1 - 20000 u_2 - 15000 u_3 = 0$$

$$-\omega^2 20 u_2 - 20000 u_1 + 45000 u_2 - 25000 u_3 = 0$$

$$-\omega^2 30 u_3 - 15000 u_1 - 25000 u_2 + 40000 u_3 = 0$$

This can be expressed in matrix notation as

$$-\omega^2 \mathbf{M}\mathbf{u} + \mathbf{K}\mathbf{u} = \mathbf{0} \qquad (2.15.3)$$

where

$$\mathbf{M} = \begin{bmatrix} 10 & 0 & 0 \\ 0 & 20 & 0 \\ 0 & 0 & 30 \end{bmatrix} \text{ kg and } \mathbf{K} = \begin{bmatrix} 45 & -20 & -15 \\ -20 & 45 & -25 \\ -15 & -25 & 40 \end{bmatrix} \text{ kN/m}$$

Equation (2.15.3) can be rearranged in a variety of ways. For example, it can be written

$$\mathbf{M}\mathbf{u} = \lambda \mathbf{K}\mathbf{u} \quad \text{where } \lambda = 1/\omega^2$$

This is an algebraic eigenvalue problem and solving it determines values for **u** and λ. MATLAB may be used to solve this problem but we do not apply it here.

Having provided an example of an eigenvalue problem we consider the standard form of this problem thus:

$$\mathbf{Ax} = \lambda\mathbf{x} \qquad\qquad (2.15.4)$$

This equation is an algebraic eigenvalue problem where **A** is a given n x n matrix of coefficients. The vector **x** is an unknown column of n elements and λ is an unknown scalar. Equation (2.15.4) can be written as

$$(\mathbf{A} - \lambda\mathbf{I})\mathbf{x} = \mathbf{0} \qquad\qquad (2.15.5)$$

Our aim is to discover the values of **x**, called the characteristic vectors or eigenvectors, and the corresponding values of λ, called the characteristic values or eigenvalues. The values of λ that satisfy (2.15.5) are given by the roots of the equation

$$|\mathbf{A} - \lambda\mathbf{I}| = 0 \qquad\qquad (2.15.6)$$

These values of λ are such that $|\mathbf{A} - \lambda\mathbf{I}|$ is singular. Since (2.15.5) is an homogeneous equation these values of λ correspond to non-trivial solutions of **x**, see section 2.5. Evaluation of the determinant in (2.15.6) leads to an nth degree polynomial in λ which is called the characteristic polynomial. This characteristic polynomial has n roots, some of which may be repeated, providing the n values of λ. Having obtained these values we can substitute back into (2.15.5) to obtain linear equations for the characteristic vectors thus:

$$(\mathbf{A} - \lambda_i\mathbf{I})\mathbf{x} = \mathbf{0} \quad i = 1, 2, ..., n \qquad\qquad (2.15.7)$$

This homogeneous system provides n non-trivial solutions for **x**. Since there are an infinity of solutions to this problem **x** must be normalised. One commonly used method of normalisation is to require that $\mathbf{x}^T\mathbf{x} = 1$.

The use of (2.15.6) and (2.15.7) is not a practical means of solving eigenvalue problems. However, MATLAB provides a function `eig` which solves this problem practically. To illustrate its use we now apply it to the problem described by Fig. 2.15.1.

```
»M=[10 0 0;0 20 0;0 0 30];
»K=1000*[45 -20 -15;-20 45 -25;-15 -25 40];
»lambda=eig(M,K)'

lambda =
    0.0073    0.0002    0.0004

»omega=sqrt([1 1 1]./lambda)

omega =
    11.7268    72.2165    52.2551
```

This tells us that the system of Fig. 2.15.1 vibrates with natural frequencies 11.72, 52.25 and 72.21 rad/s.

So far we have implicitly assumed that the coefficient matrix **A** is symmetric. If this is not the case then we consider two related eigenvalue problems, as follows:

$$\mathbf{A}\mathbf{x} = \lambda\mathbf{x}$$

and

$$\mathbf{A}^T\mathbf{y} = \beta\mathbf{y} \ \text{ or } \ \mathbf{y}^T\mathbf{A} = \beta\mathbf{y}^T$$

The equations $|\mathbf{A} - \lambda\mathbf{I}| = 0$ and $|\mathbf{A}^T - \beta\mathbf{I}| = 0$ must have the same solutions for λ and β because the determinant of a matrix and its transpose are equal. Thus the eigenvalues of these two equations are identical. However, the eigenvectors **x** and **y** may be different. The vector **x** is called the right-hand solution and **y** the left-hand solution.

We now discuss the MATLAB function eig in some detail.

2.16 THE MATLAB FUNCTION eig

There are many algorithms available to solve the eigenvalue problem. The method chosen is influenced by many factors such as the form and size of the eigenvalue problem, whether or not it is symmetric, whether it is real or complex, whether or not only the eigenvalues are required, whether all or only some of the eigenvalues and vectors are required.

We now describe the algorithms that are used in the MATLAB function eig. This MATLAB function can be used in several forms and, in the process, makes use of different algorithms. The different forms are as follows:

(1) lambda=eig(a)

(2) [u,lambda]=eig(a)

(3) lambda=eig(a,b)

(4) [u,lambda]=eig(a,b)

where lambda is a vector of eigenvalues in (1) and (3) and a diagonal matrix with the eigenvalues on the diagonal in (2) and (4). In these latter cases u is a matrix, the columns of which are the eigenvectors.

For real matrices the MATLAB function eig(a) proceeds as follows. If **A** is a general matrix, it is first reduced to Hessenberg form using Householder's transformation method. A Hessenberg matrix has zeros everywhere below the diagonal except for the first sub-diagonal. If **A** is a symmetric matrix the transform creates a tridiagonal matrix. Then the eigenvalues and eigenvectors of the real upper Hessenberg matrix are found by the iterative application of the QR procedure. The QR procedure involves decomposing the Hessenberg matrix into an upper triangular and a unitary matrix. The method is as follows:

(1) $k = 0$.

(2) Decompose $\mathbf{H}_k$ into $\mathbf{Q}_k$ and $\mathbf{R}_k$ such $\mathbf{H}_k = \mathbf{Q}_k \mathbf{R}_k$ where $\mathbf{H}_k$ is a Hessenberg or tridiagonal matrix.

(3) Compute $\mathbf{H}_{k+1} = \mathbf{R}_k \mathbf{Q}_k$. The estimates of the eigenvalues equal diag($\mathbf{H}_{k+1}$).

(4) Check the accuracy of the eigenvalues. If the process has not converged, $k = k + 1$; repeat from (2).

The values on the leading diagonal of $\mathbf{H}_k$ tend to the eigenvalues. The following script uses the MATLAB function **hess** to convert the original matrix to the Hessenberg form, followed by the iterative application of the **qr** function to determine the eigenvalues of a symmetric matrix. Note that in this script we have iterated 10 times, rather than using a formal test for convergence since the purpose of the script is merely to illustrate the functioning of the iterative application of the QR procedure.

```
a=[5 4 1 1;4 5 1 1; 1 1 4 2;1 1 2 4];
h1=hess(a);
for i=1:10
  [q r]=qr(h1);
  h2=r*q;
  h1=h2;
  p=diag(h1)';
  fprintf('%2.0f%8.4f%8.4f',i,p(1),p(2));
  fprintf('%8.4f%8.4f\n',p(3),p(4));
end
```

Running this script gives

```
 1  1.0000  8.3636  6.2420  2.3944
 2  1.0000  9.4940  5.4433  2.0627
 3  1.0000  9.8646  5.1255  2.0099
 4  1.0000  9.9655  5.0329  2.0016
 5  1.0000  9.9913  5.0084  2.0003
 6  1.0000  9.9978  5.0021  2.0000
 7  1.0000  9.9995  5.0005  2.0000
 8  1.0000  9.9999  5.0001  2.0000
 9  1.0000 10.0000  5.0000  2.0000
10  1.0000 10.0000  5.0000  2.0000
```

The iteration converges to the values 1, 10, 5 and 2 which are the correct values. This QR iteration could be applied directly to the full matrix **A** but in general it would be inefficient. We have not given details of how the eigenvectors are computed.

When there are two real or complex arguments the QZ algorithm is used instead of the QR algorithm. The QZ algorithm (Golub and Van Loan, 1989) has been modified to deal with the complex case. When **eig** is called using a single complex matrix **a** then the algorithm works by applying the QZ algorithm to **eig(a,eye(size(a)))**. The QZ algorithm begins by noting that there exists a unitary $\mathbf{Q}$ and $\mathbf{Z}$ such that $\mathbf{Q^H A Z = T}$ and $\mathbf{Q^H B Z = S}$ are both upper triangular. This is called generalised Schur decomposition. Providing s_{kk} is not zero, then the eigenvalues are computed from the ratio t_{kk}/s_{kk}, where $k = 1, 2, ..., n$. The following script demonstrates that the ratios of the diagonal elements of the $\mathbf{T}$ and $\mathbf{S}$ matrices give the required eigenvalues.

```
a=[10+2*i 1 2;1-3*i 2 -1;1 1 2];
b=[1 2-2*i -2;4 5 6;7+3*i 9 9];
[T,S,q,z,v]=qz(a,b);
r=diag(T)./diag(S)
eig(a,b)
```

Running this script gives

```
r =
   1.6154 + 2.7252i
  -0.4882 - 1.3680i
   0.1518 + 0.0193i

ans =
   1.6154 + 2.7252i
  -0.4882 - 1.3680i
   0.1518 + 0.0193i
```

Schur decomposition is closely related to the eigenvalue problem. The MATLAB function schur(a) produces an upper triangular matrix $\mathbf{T}$ with real eigenvalues on its diagonal and complex eigenvalues in 2 x 2 blocks on the diagonal. Thus $\mathbf{A}$ can be written

$$\mathbf{A = U T U^H}$$

where $\mathbf{U}$ is a unitary matrix such that $\mathbf{U^H U = I}$. The following script shows the similarity between Schur decomposition and the eigenvalues of a given matrix.

```
a=[4 -5 0 3;0 4 -3 -5;5 -3 4 0;3 0 5 4];
t=schur(a)
eig(a)
```

Running this script gives

```
t =
   12.0000   -0.0000   -0.0000   -0.0000
         0    1.0000   -5.0000   -0.0000
         0    5.0000    1.0000   -0.0000
         0         0         0    2.0000

ans =
   12.0000
    1.0000 + 5.0000i
    1.0000 - 5.0000i
    2.0000
```

We can readily identify the four eigenvalues in the matrix t.

The following script compares the performance of the eig function when solving various classes of problem.

```
disp('  real1    realsym1   real2    realsym2    comp1     comp2')
for n= 10:10:50
a=rand(n);c=rand(n);s=a+c*i;t=rand(n)+i*rand(n);
tic;[u,v]=eig(a);t1=toc;
b=a+a';d=c+c';
tic;[u,v]=eig(b);t2=toc;
tic;[u,v]=eig(a,c);t3=toc;
tic;[u,v]=eig(b,d);t4=toc;
tic;[u,v]=eig(s);t5=toc;
tic;[u,v]=eig(s,t);t6=toc;
fprintf('%8.2f%10.2f%10.2f%10.2f%10.2f%10.2f\n',t1,t2,t3,t4,t5,t6);
end;
```

This script gives the following output which shows the times taken to solve the selected eigenvalue problems.

» real1	realsym1	real2	realsym2	comp1	comp2
0.72	0.30	2.57	2.52	1.93	2.80
2.85	1.38	14.67	14.08	10.03	15.70
7.37	3.70	43.30	41.93	30.88	45.22
17.15	7.52	93.67	84.42	66.63	100.62
30.30	13.60	175.85	166.97	124.18	175.45

The above script can easily be modified to count the number of flops/10^4, rather than measuring computing time. If this is done the output is as follows:

» real1	realsym1	real2	realsym2	comp1	comp2
2.86	0.98	15.77	17.27	12.81	18.69
24.23	7.75	144.82	124.96	100.58	149.97
79.20	27.18	433.11	420.80	334.17	485.70
170.83	63.90	1063.06	1052.92	798.56	1110.32
321.44	121.40	2047.58	2030.45	1526.15	2195.80

In some circumstances not all the eigenvalues and eigenvectors are required. For example, in a complex engineering structure, modelled with many hundreds of degrees of freedom, we may only require the first 15 eigenvalues, giving the natural frequencies of the model, and the corresponding eigenvectors. MATLAB does not provide any algorithms that take advantage of this fact to speed up calculations. Either the reader must write a script to implement an efficient algorithm, or an eigenvalue reduction algorithm should be used to reduce the size of the original eigenvalue problem (Guyan, 1965).

MATLAB also includes the facility to find the eigenvalues of a sparse matrix. The following script compares the time taken to find the eigenvalues of a matrix treated as sparse with the corresponding time taken to find the eigenvalues of the corresponding full matrix.

```
%generate a sparse triple diagonal matrix
rowpos=2:100;colpos=1:99;
values=ones(1,99);
offdiag=sparse(rowpos,colpos,values,100,100);
a=sparse(1:100,1:100,4*ones(1,100),100,100);
a=a+offdiag+offdiag';
%generate full matrix
b=full(a);
tic;eig(a);t1=toc;
tic;eig(b);t2=toc;
fprintf('time sparse eigen solve= %4.2f\n',t1);
fprintf('time full eigen solve= %4.2f\n',t2);
```

The results from running this script are

```
»time sparse eigen solve= 9.48
time full eigen solve= 22.48
```

Clearly a significant saving in time.

2.17 SUMMARY

We have described many of the important algorithms related to computational matrix algebra and shown how the power of MATLAB can be used to illustrate the application of these algorithms in a revealing way. The scripts provided should help the reader to develop applications.

PROBLEMS

2.1. Write MATLAB scripts to implement both the Gauss–Seidel and the Jacobi method, and use them to solve, with an accuracy of 0.000005, the equation system $Ax = b$ where the elements of A are

$$a_{ii} = -4 \text{ and } a_{ij} = 2 \text{ if } |i - j| = 1$$
$$= 0 \text{ if } |i - j| \geq 2 \quad \text{where } i, j = 1, 2, ..., 10$$

and

$$\mathbf{b}^{\mathrm{T}} = [2 \quad 3 \quad 4 ... 11]$$

Check your results by solving the system using the \ operator.

2.2. An $n \times n$ Hilbert matrix, A, is defined by

$$a_{ij} = 1/(i + j - 1) \text{ for } i, j = 1, 2, ..., n$$

Find the inverse of A and the inverse of $A^{\mathrm{T}}A$ for $n = 5$. Then, noting that

$(\mathbf{A}^T\mathbf{A})^{-1} = \mathbf{A}^{-1}(\mathbf{A}^{-1})^T$

find the inverse of $\mathbf{A}^T\mathbf{A}$ using this result for values of $n = 3, 4, ..., 6$. Compare the accuracy of the two results by using the inverse Hilbert function `invhilb` to find the exact inverse using $(\mathbf{A}^T\mathbf{A})^{-1} = \mathbf{A}^{-1}(\mathbf{A}^{-1})^T$. *Hint*: compute norm$(\mathbf{P} - \mathbf{R})$ and norm$(\mathbf{Q} - \mathbf{R})$ where $\mathbf{P} = (\mathbf{A}^T\mathbf{A})^{-1}$ and $\mathbf{Q} = \mathbf{A}^{-1}(\mathbf{A}^{-1})^T$ and $\mathbf{R}$ is the exact inverse.

2.3. Find the condition number of $\mathbf{A}^T\mathbf{A}$ where $\mathbf{A}$ is an n x n Hilbert matrix, defined in problem 2.2, for $n = 3, 4, ..., 6$. How do these results relate to the results of problem 2.2?

2.4. It can be proved that the series $(\mathbf{I} - \mathbf{A})^{-1} = \mathbf{I} + \mathbf{A} + \mathbf{A}^2 + \mathbf{A}^3 + ...$, where $\mathbf{A}$ is an n x n matrix, converges if the eigenvalues of $\mathbf{A}$ are all less than unity. The following n x n matrix satisfies this condition if $a + 2b < 1$:

$$\begin{bmatrix} a & b & 0 & \cdots & 0 & 0 & 0 \\ b & a & b & \cdots & 0 & 0 & 0 \\ \vdots & \vdots & \vdots & & \vdots & \vdots & \vdots \\ 0 & 0 & 0 & \cdots & b & a & b \\ 0 & 0 & 0 & \cdots & 0 & b & a \end{bmatrix}$$

Experiment with this matrix for various values of n, a and b to illustrate that the series converges under the condition stated.

2.5. Use the function `eig` to find the eigenvalues of the following matrix:

$$\begin{bmatrix} 2 & 3 & 6 \\ 2 & 3 & -4 \\ 6 & 11 & 4 \end{bmatrix}$$

Then use the `rref` function on the matrix $(\mathbf{A} - \lambda\mathbf{I})$, taking λ equal to any of the eigenvalues. Solve the resulting equations by hand to obtain the eigenvector of the matrix, by noting that the eigenvectors are the solution of $(\mathbf{A} - \lambda\mathbf{I})\mathbf{x} = \mathbf{0}$ for λ equal to the eigenvalues, assuming an arbitrary value for x_3.

2.6. Solve the system given in problem 2.4 for $n = 20:10:50$, treating it as both a full and a sparse matrix. Let $\mathbf{b} = [1\ 1\ ...\ 1]$, $a = 2$ and $b = 1$. Compare the flops taken to solve the full and sparse systems.

2.7. For the system given in problem 2.4, find the eigenvalues, assuming both full and sparse forms with $n = 10:10:30$. Compare your results with the exact solution given by

$$\lambda_k = a + 2b\ \cos\{k\pi/(n + 1)\}, \, k = 1, 2, ...$$

2.8. Use the function `sprandsym(10,0.2,r)` to generate a 10 x 10 positive definite
      sparse matrix with density 0.2. By taking `r = 3:2:21` the matrix will be positive
      definite. Find the number of flops required for the Cholesky decomposition of this
      matrix. Convert to full form and repeat the experiment.

2.9. Find the solution of the over-determined system given below using `pinv`, `qr` and
      the $\backslash$ operator.

$$
\begin{bmatrix}
2 & -3 & 2 \\
1.9 & -3 & 2.2 \\
2.1 & -2.9 & 2 \\
6.1 & 2.1 & -3 \\
-3 & 5 & 2.1
\end{bmatrix}
\begin{bmatrix} x_1 \\ x_2 \\ x_3 \end{bmatrix}
=
\begin{bmatrix}
1.01 \\
1.01 \\
0.98 \\
4.94 \\
4.10
\end{bmatrix}
$$

2.10. Write a script to generate $\mathbf{E} = \{1/(n + 1)\}\mathbf{C}$ where

$$
\begin{aligned}
c_{ij} &= i(n - i + 1) & \text{if } i = j \\
&= c_{i,j-1} - i & \text{if } j > i \\
&= c_{ji} & \text{if } j < i
\end{aligned}
$$

Having generated $\mathbf{E}$ solve $\mathbf{Ex} = \mathbf{b}$ where $\mathbf{b} = [1: n]^T$ by

      (a) using the $\backslash$ operator;
      (b) using the `lu` function and solving $\mathbf{Ux} = \mathbf{y}$ and $\mathbf{Ly} = \mathbf{b}$.

2.11. Determine the inverse of $\mathbf{E}$ of problem 2.10 for $n = 20$ and 50. Compare with the
      exact inverse which is a matrix with 2 along the main diagonal and -1 along the
      upper and lower sub-diagonals and zero elsewhere.

2.12. Determine the eigenvalues of $\mathbf{E}$ defined in problem 2.10 above for $n = 20$ and 50.
      The exact eigenvalues for this system are given by $\lambda_k = 1/[2 - 2\cos\{k\pi/(n + 1)\}]$
      where $k = 1, ..., n$.

2.13. Determine the condition number of $\mathbf{E}$ of problem 2.10 using the MATLAB function
      cond, for $n = 20$ and 50. Compare your results with the theoretical expression
      for the condition number which is $4n^2/\pi^2$.

2.14. Find the eigenvalues and the left and right eigenvectors using the MATLAB function
      `eig` for the matrix

$$
\mathbf{A} =
\begin{bmatrix}
8 & -1 & -5 \\
-4 & 4 & -2 \\
18 & -5 & -7
\end{bmatrix}
$$

3

Roots of equations

3.1 INTRODUCTION

The problem of solving non-linear equations arises frequently and naturally from the study of a wide range of practical problems. The problem may involve a system of non-linear equations in many variables or one equation in one unknown. We shall initially confine ourselves to considering the solution of one equation in one unknown. The general form of the problem may be simply stated as finding a value of the variable x such that

$$f(x) = 0$$

where f is any non-linear function of x. The value of x is then called a solution or root of this equation and may be just one of many solutions.

To illustrate our discussion and provide a practical insight into the solution of non-linear equations we shall consider an equation described by Armstrong and Kulesza (1981). These authors report a problem which arises from the study of resistive mixer circuits. Given an applied current and voltage it is necessary to find the current flowing in part of the circuit. This leads to a simple non-linear equation which after some manipulation may be expressed in the form

$$x \exp(-x/c) = 0 \text{ or equivalently } x = \exp(-x/c) \qquad (3.1.1)$$

Here c is a given constant and x the variable we wish to determine. The solution of such equations is not obvious but Armstrong and Kulesza provide an approximate solution based on a series expansion which gives a reasonably accurate solution of this equation for a large range of values of c. This approximation is given in terms of c by

$$x = cu[1 - \log_e\{(1 + c)u\}/(1 + u)] \qquad (3.1.2)$$

where $u = \log_e(1 + 1/c)$. This is an interesting and useful result since it is reasonably accurate for values of c in the five decade range $[10^{-3}, 100]$ and gives a relatively easy way of finding the solutions of a whole family of equations generated by varying c. Although this result is useful for this particular equation, when we attempt to use this type of *ad hoc* approach for the general solution of non-linear equations there are significant drawbacks. These are:

(1) *ad hoc* approaches to the solutions of equations are rarely as successful as this example in finding a formula for the solution of a given equation; usually it is impossible to obtain such formulae;

(2) even when they exist such formulae require considerable time and ingenuity to develop;

(3) we may require greater accuracy than any *ad hoc* formula can provide.

To illustrate point (3) consider Fig. 3.1.1 which is generated by the MATLAB script below. This figure shows the results obtained using the formula (3.1.2) together with the results using the MATLAB toolbox function `fzero` to solve the non-linear equation (3.1.1).

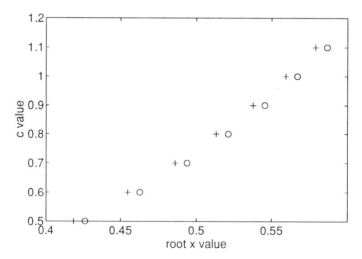

Fig. 3.1.1. Solution of $x = \exp(-x/c)$. Results from the MATLAB function `fzero` are indicated by a "o" and those from the Armstrong and Kulesza formula by a "+".

```
global c1
ro=[ ];ve=[ ];x=[ ];
c=.5:.1:1.1;
u=log(1+1 ./c);
x=c.*u.*(1-log((1+c).*u)./(1+u));
%Now solve equation using MATLAB function fzero
i=0;
for c1=.5:.1:1.1
  i=i+1;
  ro(i)=fzero('f301',1,0.00005);
end;
plot(x,c,'+');
axis([.4 .6 .5 1.2]);
hold on
plot(ro,c,'o');
xlabel('root x value'); ylabel('c value');
hold off
```

The function `fzero` is discussed in detail in section 3.10. Note that the call of `fzero` takes the form: `fzero('f301',1,0.00005)`. This gives an accuracy of 0.00005 for the roots and uses an initial approximation 1. The function `fzero` provides the root with up to 16 digit accuracy, if required, whereas the formula (3.1.2) of Armstrong and Kulesza, although faster, gives the result to one or two decimal places only. In fact the method of Armstrong and Kulesza becomes more accurate for large values of c. The function script `f301` is defined thus:

```
function fv=f301(x)
global c1
fv=x-exp(-x/c1);
```

The variable `c1` is declared as a global variable in both the function `f301` and the script where the function is called, so that values of `c1` are passed to the function definition. This allows us to use a function definition consistent with the requirements of `fzero`.

From the above discussion we conclude that although occasionally ingenious alternatives may be available, in the vast majority of cases we must use algorithms which provide, with reasonable computational effort, the solutions of general problems to any specified accuracy. Before describing the nature of these algorithms in detail we consider different types of equations and the general nature of their solutions.

3.2 THE NATURE OF SOLUTIONS TO NON-LINEAR EQUATIONS

We illustrate the nature of the solutions to non-linear equations by considering two examples which we wish to solve for the variable x.

(1) $(x - 1)^3(x + 2)^2(x - 3) = 0$,

$$\text{i.e. } x^6 - 2x^5 - 8x^4 + 14x^3 + 11x^2 - 28x + 12 = 0$$

(2) $\exp(-x/10)\sin(10x) = 0$

Example (1) is a special type of non-linear equation known as a polynomial equation since it involves only integer powers of the variable x and no other function. Such polynomial equations have the important characteristic that they have n roots where n is the degree of the polynomial. In example (1) the highest power of x, and hence the degree of the polynomial, is six. The solutions of a polynomial may be complex or real, separate or coincident. Fig. 3.2.1 illustrates the nature of the solutions of example (1). Although there must be six roots, there are three coincident roots at $x = 1$ and two coincident roots at $x = -2$. There is also a single root at $x = 3$. Coincident roots may present difficulties for some algorithms as do roots which are very close together so it is important to appreciate their existence. The user may require a particular root of the equation or all the roots. In the case of polynomial equations special algorithms exist to find all the roots.

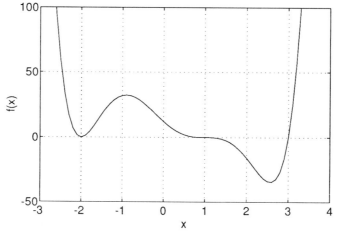

Fig. 3.2.1. Plot of the function $f(x) = (x - 1)^3(x + 2)^2(x - 3)$.

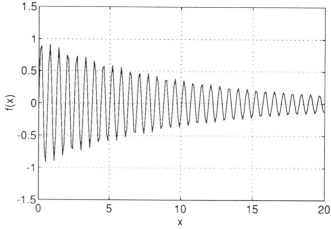

Fig. 3.2.2. Plot of $f(x) = \exp(-x/10) \sin(10x)$.

For non-linear equations involving transcendental functions the task of finding all the roots is a daunting one, since the number of roots may not be known or there may be an infinity of roots. This situation is illustrated by Fig. 3.2.2 which shows the graph of example (2) for x in the range $(0, 20)$. If we extended the range of x more roots would be revealed.

We shall now consider some simple algorithms to find a specific root of a given non-linear equation.

3.3 THE BISECTION ALGORITHM

This simple algorithm assumes that an initial interval is known in which a root of the equation $f(x) = 0$ lies and then proceeds to reduce this interval until the required accuracy is achieved for the root. This algorithm is mentioned only briefly since it is not in practice used by itself but in conjunction with other algorithms to improve their reliability. The algorithm may be described by:

input   interval in which the root lies
while   interval too large
   (1) Bisect the current interval in which the root lies.
   (2) Determine in which half of the interval the root lies.
end
display (root)

The principles on which this algorithm are based are simple. Given an initial interval in which a specific root lies the algorithm will provide an improved approximation for the root. However, the requirement that an interval is known is sometimes difficult to achieve and although the algorithm is reliable it is extremely slow.

Alternative algorithms have been developed which converge more rapidly and this chapter is concerned with describing some of the most important of these. All the algorithms we consider are iterative in character, i.e. they proceed by repeating the same sequence of steps until the root approximation is accurate enough to satisfy the user. We now consider the general form of an iterative method, the nature of the convergence of such methods and the problems they encounter.

3.4 ITERATIVE OR FIXED POINT METHODS

We require to solve the general equation $f(x) = 0$ but to illustrate iterative methods clearly we consider a simple example. Suppose we wish to solve the quadratic

$$x^2 - x - 1 = 0 \qquad\qquad (3.4.1)$$

This equation can be solved by using the standard formula for solving quadratics but we take a different approach. Rearrange (3.4.1) as follows:

$$x = 1 + 1/x$$

Then rewrite it in iterative form using subscripts as follows:

$$x_{r+1} = 1 + 1/x_r \text{ for } r = 0, 1, 2, \ldots \tag{3.4.2}$$

Assuming we have an initial approximation x_0 to the root we are seeking, we can proceed from one approximation to another using this formula. The iterates we obtain in this way may or may not converge to the solution of the original equation. This is not the only iterative procedure for attempting to solve (3.4.1); we can generate two others from (3.4.1) as follows:

$$x_{r+1} = x_r^2 - 1 \text{ for } r = 0, 1, 2, \ldots \tag{3.4.3}$$

and

$$x_{r+1} = \sqrt{(1 + x_r)} \text{ for } r = 0, 1, 2, \ldots \tag{3.4.4}$$

Starting from the same initial approximation, these iterative procedures may or may not converge to the same root. Table 3.4.1 shows what happens when we use the initial approximation $x_0 = 2$ with the iterative procedures (3.4.2), (3.4.3) and (3.4.4). It shows that iterations (3.4.2) and (3.4.4) converge but (3.4.3) does not.

Note that when the root is reached no further improvement is possible and the point remains fixed. Hence the roots of the equation are the *fixed points* of the iteration. To remove the unpredictability of this approach we must be able to find general conditions which determine when such iterative schemes converge, when they do not and the nature of this convergence.

Table 3.4.1. Difference between exact root and iterate for $x^2 - x - 1 = 0$.

Iteration (3.4.2)	Iteration (3.4.3)	Iteration (3.4.4)
−0.1180	1.3820	0.1140
0.0486	6.3820	0.0349
−0.0180	61.3820	0.0107
0.0070	3966.3820	0.0033
−0.0026	15745021.3820	0.0010

3.5 THE CONVERGENCE OF ITERATIVE METHODS

The procedure described in section 3.4 can be applied to any equation $f(x) = 0$ and has the general form

$$x_{r+1} = g(x_r) \quad r = 0, 1, 2, \ldots \tag{3.5.1}$$

It is not our purpose to give the details of the derivation of convergence conditions for this form of iteration but to point out some of the difficulties which may arise in using them even when this condition is satisfied. The detailed derivation is given in many text books, see for example Lindfield and Penny (1989). It can be shown that the approximate relation between the current error ε_{r+1} at the $(r + 1)$th iteration and the previous error ε_r is given by

$$\varepsilon_{r+1} = \varepsilon_r g'(t_r)$$

where t_r is a point lying between the exact root and the current approximation to the root. Thus the error will be decreasing if the absolute value of the derivative at these points is less than 1. However, this does not guarantee convergence from all starting points and the initial approximation must be sufficiently close to the root for convergence to occur.

In the case of the specific iterative procedures (3.4.2) and (3.4.3), Table 3.5.1 shows how the values of the derivatives of the corresponding $g(x)$ vary with the values of the approximations to x_r. This table provides numerical evidence for the theoretical assertion in the case of iterations (3.4.2) and (3.4.3).

Table 3.5.1. The values of the derivatives
for iterations given by (3.4.2) and (3.4.3).

Iteration (3.4.2)	derivative	Iteration (3.4.3)	derivative
−0.1180	0.44	1.3820	6.00
0.0486	0.36	6.3820	16.00
−0.0180	0.39	61.3820	126.00
0.0070	0.38	3966.3820	7936.00

However, the concept of convergence is more complex than this. We need to give some answer to the crucial question: if an iterative procedure converges how can we classify the rate of convergence? We will not derive this result but refer the reader to Lindfield and Penny (1989) and state the answer to the question. Suppose all derivatives of the function $g(x)$ of order 1 to $p - 1$ are zero at the exact root a. Then the relation between the current error ε_{r+1} at the $(r + 1)$th iteration and the previous error ε_r is given by

$$\varepsilon_{r+1} = (\varepsilon_r)^p g^{(p)}(t_r)/p! \qquad (3.5.2)$$

where t_r lies between the exact root and the current approximation to the root and $g^{(p)}$ denotes the pth derivative of g. The importance of this result is that it means the current error is proportional to the pth power of the previous error and clearly, on the basis of the reasonable assumption that the errors are much smaller than 1, the higher the value of p the faster the convergence. Such methods are said to have pth order convergence. In general it is cumbersome to derive iterative methods of order higher than two or three and second-order methods have proved very satisfactory in practice for solving a wide range of non-linear equations. In this case the current error is proportional to the square of the previous error. This is often called quadratic convergence; if the error is proportional to the previous error it is called linear convergence. This provides a convenient classification for the convergence of iterative methods but avoids the difficult questions: for what range of starting values will the process converge and how sensitive is convergence to changes in the starting values?

3.6 RANGES FOR CONVERGENCE AND CHAOTIC BEHAVIOUR

We illustrate some of the problems of convergence by considering a specific example which highlights some of the difficulties. Short (1992) examined the behaviour of the iterative process

$$x_{r+1} = -0.5(x_r^3 - 6x_r^2 + 9x_r - 6) \text{ for } r = 0, 1, 2, \dots$$

for solving the equation $(x - 1)(x - 2)(x - 3) = 0$. This iterative procedure clearly has the form

$$x_{r+1} = g(x_r) \quad r = 0, 1, 2, \dots$$

and it is easy to verify it has the following properties:

$$g'(1) = 0 \quad \text{and} \quad g''(1) \neq 0$$
$$g'(2) \neq 0$$
$$g'(3) = 0 \quad \text{and} \quad g''(3) \neq 0.$$

Thus by taking $p = 2$ in result (3.5.2) we can expect, for appropriate starting values, quadratic convergence for the roots at $x = 1$ and $x = 3$ but at best linear convergence for the root at $x = 2$. The major problem is, however, to determine the ranges of initial approximation which will converge to the different roots. This is not an easy task but one simple way of doing this is to draw a graph of $y = x$ and $y = g(x)$. The points of intersection provide the roots. The line $y = x$ has a slope of 1 and points where the slope of $g(x)$ is less than this provide a range of initial approximations which converge to one or other of the roots.

This graphical analysis shows that points within the range 1 to 1.43 (approximately) converge to the root 1 and points in the range 2.57 (approximately) to 3 converge to the root 3. This is the obvious part of the analysis. However, Short demonstrates that there are many other ranges of convergence for this iterative procedure, many of them very narrow indeed, which lead to chaotic behaviour in the iterative process. He demonstrates for example that taking $x_0 = 4.236067968$ will converge to the root $x = 3$ whereas taking $x_0 = 4.236067970$ converges to the root $x = 1$, a remarkable change for such a small variation in the initial approximation. This should serve as a warning to the reader that the study of convergence properties is in general not an easy task.

Fig. 3.6.1 illustrates this point quite strikingly. It shows the graph of x and the graph of $g(x)$ where

$$g(x) = -0.5(x^3 - 6x^2 + 9x - 6)$$

The x line intersects with $g(x)$ to give the roots of the original equation. The graph also

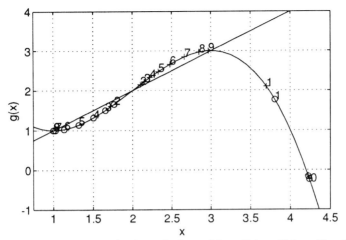

Fig. 3.6.1. Iterates in the solution of $(x-1)(x-2)(x-3) = 0$
from close but different starting points.

shows iterates starting from $x_0 = 4.236067968$, indicated by a "o", and iterates starting from $x_0 = 4.236067970$, indicated by a "+". The starting points are so close they are of course superimposed on the graph. However the iterates soon take their separate paths to converge on different roots of the equation. The path indicated by "o" converges to the root $x = 3$ and the path indicated by "+" converges to the root $x = 1$. The sequence of numbers on the graph shows the last nine iterates. The point referenced by zero is in fact all the points which are initially very close together. This is a remarkable example and users should verify these phenomena for themselves by running the following MATLAB script.

```
x=.75:.1:4.5; g=-0.5*(x.^3-6*x.^2+9*x-6);
plot(x,g); axis([.75,4.5,-1,4]);
hold on; plot(x,x);
xlabel('x'); ylabel('g(x)'); grid on;
ch=['o','+']; ty=0;
num=[ '0','1','2','3','4','5','6','7','8','9'];
for x1=[4.236067970  4.236067968]
  ty=ty+1;
  for i=1:19
    x2=-0.5*(x1^3-6*x1^2+9*x1-6);
    %First ten points very close, so represent by '0'
    if i==10,
      text(4.25,-.2,'0');
    elseif i>10
      text(x1,x2+.1,num(i-9));
    end;
    plot(x1,x2,ch(ty)); x1=x2;
  end;
end;
hold off
```

It is interesting to note that the iterative form

$$x_{r+1} = x_r^2 + c \ \text{ for } \ r = 0, 1, 2, \ldots$$

demonstrates strikingly chaotic behaviour when the iterates are plotted in the complex plane and for complex ranges of values for c.

We now return to the more mundane task of developing algorithms that work in general for the solution of non-linear equations. In the next section we shall consider a simple method of order 2.

3.7 NEWTON'S METHOD

This method for the solution of the equation $f(x) = 0$ is based on the simple geometric properties of the tangent to the curve $f(x)$. The method requires some initial approximation to the root and that the derivative of $f(x)$ exists in the range of interest. Fig. 3.7.1 illustrates the operation of the method. The diagram shows the tangent to the curve at the current approximation x_0. This tangent strikes the x-axis at x_1 and provides us with an improved approximation to the root. Similarly the tangent at x_1 gives the improved approximation x_2. The process is repeated until some convergence criterion is satisfied. It is easy to translate this geometrical procedure into a numerical method for finding the root since the tangent of the angle between the x-axis and the tangent equals

$$f(x_0)/(x_1 - x_0)$$

and the slope of this tangent itself equals $f'(x_0)$, the derivative of $f(x)$ at x_0. So we have the equation

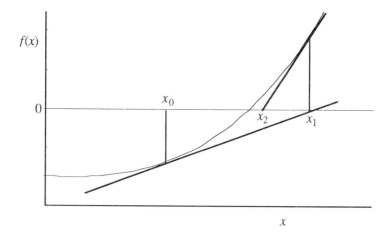

Fig. 3.7.1. Geometric interpretation of Newton's method.

$$f'(x_0) = f(x_0)/(x_1 - x_0)$$

Thus the improved approximation, x_1, is given by

$$x_1 = x_0 - f(x_0)/f'(x_0)$$

This may be written in iterative form as

$$x_{r+1} = x_r - f(x_r)/f'(x_r) \quad \text{where } r = 0, 1, 2, \ldots \tag{3.7.1}$$

We note that this method is of the general iterative form

$$x_{r+1} = g(x_r) \quad \text{where } r = 0, 1, 2, \ldots$$

Consequently the discussion of section 3.5 applies to it. On computing $g'(a)$, where a is the exact root, we find it is zero. However, $g''(a)$ is in general non-zero so the method is of order 2 and we will expect convergence to be quadratic. For a sufficiently close initial approximation convergence to the root will be rapid.

A MATLAB function fnewton is supplied for Newton's method. The function that forms the left-hand side of the equation we wish to solve *and* its derivative must be supplied by the user as function scripts; these are given as the first and second parameters of the function. The third parameter is an initial approximation to the root. The convergence criterion used is that the difference between successive approximations to the root is less than a small preset value. This value must be supplied by the user and is given as the fourth parameter of the function.

```
function [res, it]=fnewton(func,dfunc,x,tol)
% x is an initial starting value, tol is required accuracy
it=0; x0=x;
d=feval(func,x0)/feval(dfunc,x0);
while abs(d)>tol
  x1=x0-d;
  it=it+1;
  x0=x1;
  d=feval(func,x0)/feval(dfunc,x0);
end;
res=x0;
```

We will now find a root of the equation

$$x^3 - 10x^2 + 29x - 20 = 0$$

To use Newton's method we must define the function and its derivative thus:

```
function F=f302(x);
F=x.^3-10.0*x.^2+29.0*x-20.0;

function F=f303(x)
F=3*x.^2-20*x+29;
```

We may call the function fnewton as follows:

```
»[x,it]=fnewton('f302','f303',7,.00005)
x =
    5.0000

it =
    6
```

The progress of the iterations when solving $x^3 - 10x^2 + 29x - 20 = 0$ by Newton's method is shown in Fig. 3.7.2.

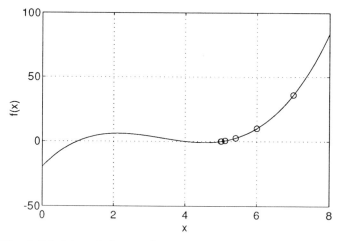

Fig. 3.7.2. Plot of $x^3 - 10x^2 + 29x - 20 = 0$ with the iterates of
Newton's method shown by a "o".

Table 3.7.1 gives numerical results for this problem when Newton's method is used to seek a root, starting the iteration at –2. The second column of the table gives the current error ε_r by subtracting the known exact root from the current iterate. The third column contains the value of $2\varepsilon_{r+1}/\varepsilon_r^2$. This value tends to a constant as the process proceeds. From theoretical considerations this value should approach the second derivative of the right-hand side of the Newton iterative formula. This follows from (3.5.2) with $p = 2$. The final column contains the value of the second-order derivative of $g(x)$ calculated as follows. From (3.7.1) we have $g(x) = x - f(x)/f'(x)$. Thus from this we have

$$g'(x) = 1 - [\{f'(x)\}^2 - f''(x)f(x)]/[f'(x)]^2 = f''(x)f(x)/[f'(x)]^2$$

On differentiating again,

$$g''(x) = [\{f'(x)\}^2\{f'''(x)f(x) + f''(x)f'(x)\} - 2f'(x)\{f''(x)\}^2 f(x)]/[f'(x)]^4$$

Putting $x = a$, where a is the exact root, since $f(a) = 0$ we have

$$g''(a) = f''(a)/f'(a) \qquad\qquad (3.7.2)$$

Thus we have a value for the second derivative of $g(x)$ when $x = a$. We note that as x approaches the root the final column of Table 3.7.1, which uses this formula, gives an increasingly accurate approximation to the second derivative of $g(x)$. Thus the table verifies our theoretical expectations.

Table 3.7.1. Newton's method to solve $x^3 - 10x^2 + 29x - 20 = 0$
with an initial approximation of -2.

x value	error ε_r	$2\varepsilon_{r+1}/\varepsilon_r^2$	Approximate 2nd derivative
-2.000000	3.000000	0.320988	0.395062
-0.444444	1.444444	0.513956	0.589028
0.463836	0.536164	0.792621	0.845266
0.886072	0.113928	1.060275	1.076987
0.993119	0.006881	1.159637	1.160775
0.999973	0.000027	1.166639	1.166643
1.000000	0.000000	1.166639	1.166667

Newton's method requires the first derivative of $f(x)$ to be supplied by the user. To make the procedure more self-contained we can use a standard approximation to the first derivative which takes the form

$$f'(x_r) = \{f(x_r) - f(x_{r-1})\}/(x_r - x_{r-1}) \qquad\qquad (3.7.3)$$

Substituting this result in (3.7.1) gives the new procedure for calculating the improvements to x as

$$x_{r+1} = \{x_{r-1} f(x_r) - x_r f(x_{r-1})\}/(x_r - x_{r-1}) \text{ for } r = 0, 1, 2, ... \qquad (3.7.4)$$

This method does not require the calculation of the first derivative of $f(x)$ but does require that we know two initial approximations to the root, x_0 and x_1. Geometrically we have simply approximated the slope of the tangent to the curve by the slope of a secant. For this reason the method is known as the *secant method*. The convergence of this method is slower than Newton's method. Another procedure similar to the secant method is called *regula falsi*. In this method two values of x which enclose the root are chosen to start the next iteration rather than the most recent pair of x values as in the secant method.

Newton's method and the secant method work well on a wide range of problems. However, for problems where the roots of an equation are close together or equal the convergence may be slow. We now consider a simple adjustment to Newton's method which provides good convergence even with multiple roots.

3.8 SCHRODER'S METHOD

In section 3.2 we described how coincident roots present significant problems for most algorithms. In the case of Newton's method its performance is no longer quadratic for finding a coincident root and the procedure must be modified if it is to maintain this property. The iteration for Schroder's method for finding multiple roots has a form similar to that of Newton's method given in (3.7.1) except for the inclusion of a multiplying factor m. Thus

$$x_{r+1} = x_r - mf(x_r)/f'(x_r) \quad \text{for } r = 0, 1, 2, \ldots \tag{3.8.1}$$

Here m is an integer equal to the multiplicity of the root to which we are trying to converge. Since the user may not know the value of m it may have to be found experimentally.

It can be verified by some simple but lengthy algebraic manipulation that for a function $f(x)$ with multiple roots at $x = a$, $g'(a) = 0$. Here $g(x)$ is the right-hand side of equation (3.8.1) and a is the exact root. This modification is sufficient to preserve the quadratic convergence of Newton's method

A MATLAB function for Schroder's method, schroder, is provided as follows:

```
function [res, it]=schroder(func,dfunc,m,x,tol)
% The function has a root of multiplicity m.
% x is a starting value, tol is required accuracy
it=0; x0=x;
d=feval(func,x0)/feval(dfunc,x0);
while abs(d)>tol
  x1=x0-m*d;
  it=it+1; x0=x1;
  d=feval(func,x0)/feval(dfunc,x0);
end;
res=x0;
```

We will now use the function schroder to solve $(e^{-x} - x)^2 = 0$. In this case we must set the multiplying factor m to 2. We use the call of the function schroder as follows:

```
»[x,it]=schroder('f304','f305',2,-2,.00005)
x =
    0.5671

it =
    5
```

where f304 and its derivative f305 are defined as follows:

```
function F=f304(x);
F=(exp(-x)-x).^2;

function F=f305(x);
F=2.0*(exp(-x)-x).*(-exp(-x)-1);
```

It is interesting to note that Newton's method took 17 iterations to solve this problem in contrast to the five required by Schroder's method.

When a function $f(x)$ is known to have repeated roots, an alternative to Schroder's approach is to apply Newton's method to the function $f(x)/f'(x)$ rather than to the function $f(x)$ itself. It can be easily shown by direct differentiation that if $f(x)$ has a root of any multiplicity then $f(x)/f'(x)$ will have the same root but with multiplicity 1. Thus the algorithm has the iterative form (3.7.1) but modified by replacing $f(x)$ with $f(x)/f'(x)$. The advantage of this approach is that the user does not have to know the multiplicity of the root which is to be found. The considerable disadvantage is that both the first- and second-order derivatives must be supplied by the user.

3.9 NUMERICAL PROBLEMS

We now consider the following problems which arise in solving single variable non-linear equations.

(1)   Finding good initial approximations.
(2)   Ill-conditioned functions.
(3)   Deciding on the most suitable convergence criteria.
(4)   Discontinuities in the equation to be solved.

These problems are now examined in detail.

(1) Finding an initial approximation can be difficult for some non-linear equations and a graph can be a considerable help in supplying such a value. The advantage of working in a MATLAB environment is that the script for the graph of the function can be easily generated and input can be taken from it directly. The function plotapp defined here finds an approximation to the root of a function supplied by the user in the range given by parameters rangelow and rangeup using a step given by interval.

```
function approx=plotapp(func,rangelow,interval,rangeup)
% Finds graphical approximations for the root of func
approx=[ ];
x=rangelow:interval:rangeup;
plot(x,feval(func,x));
hold on;
xlabel('x'); ylabel('f(x)');
```

[Script continues ...

```
title(' ** Place cursor close to root and click mouse ** ')
grid on;
%Use ginput to get approximation from graph using mouse
approx=ginput(1);
fprintf('Approximate root is %8.2f\n',approx(1));
hold off
```

The script below shows how this function may be used with the MATLAB function fzero to find a root of $x - \cos x = 0$.

```
approx = plotapp('f306',-2,0.1,2);
% Use this approximation use fzero to find exact root
root=fzero('f306',approx(1),0.00005);
fprintf('Exact root is %8.5f\n',root);
```

The function f306 is defined as follows:

```
function fv=f306(x);
fv=x-cos(x);
```

Fig. 3.9.1 gives the graph of the function f306 generated by plotapp and shows the cross-hair cursor generated by the ginput function close to the root. The call ginput(1) means only one point is taken. The cursor can be positioned over the intersection of the curve with the x-axis. This provides a useful initial approximation, the accuracy of which depends on the scale of the graph. In the case of this example an initial approximation was found to be 0.74 and the exact value was found using fzero to be 0.7391.

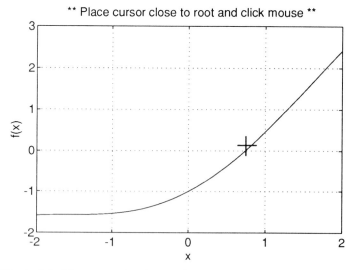

Fig. 3.9.1. The cursor is shown close to the position of the root.

(2) Ill-conditioning in a non-linear equation means that small changes in the coefficients of the equation lead to unexpectedly large errors in the solutions. An interesting example of a very ill-conditioned polynomial is Wilkinson's polynomial. The MATLAB function poly(v) generates the coefficients of a polynomial, beginning with the coefficient of the highest power, with roots which are equal to the elements of the vector v. Thus poly(1:n) generates the coefficients of the polynomial with the roots 1, 2, ..., n which is Wilkinson's polynomial of degree $n - 1$.

(3) In the design of any numerical algorithm for the solution of non-linear equations the termination criterion is particularly important. There are two major indicators of convergence: the difference between successive iterates and the value of the function at the current iterate. Taken separately these indicators may be misleading. For example, some non-linear functions are such that small changes in the independent variable value may lead to large changes in the function value. In this case it may be better to monitor both indicators.

(4) The function $f(x) = \sin(1/x)$ is particularly difficult to plot and the equation $\sin(1/x) = 0$ very difficult to solve since it has an infinite number of roots, all clustered between 1 and −1. The function has a discontinuity at $x = 0$. Fig. 3.9.2 *attempts* to illustrate the behaviour of this function. In fact the graph shown does not truly represent the function and this plotting problem is discussed in more detail in Chapter 4. Near a discontinuity the function changes rapidly for small changes in the independent variable and some algorithms may have problems with this.

All the above points emphasise the need for algorithms for solving non-linear equations to be not only fast and efficient but robust as well. The next algorithm combines these properties and is relatively undemanding on the user.

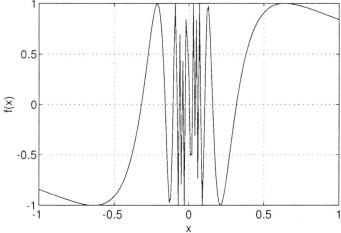

Fig. 3.9.2. Plot of $f(x) = \sin(1/x)$. This plot is spurious in the range ±0.2.

3.10 THE MATLAB FUNCTION fzero AND COMPARATIVE STUDIES

Some problems may present particular difficulties for algorithms which in general work well. For example, algorithms which have fast ultimate convergence may initially diverge. One way to improve the reliability of an algorithm is to ensure that at each stage the root is confined to a known interval and the method of bisection, introduced in section 3.3, may be used to provide an interval in which the root lies. Thus a method which combines bisection with a rapidly convergent procedure may be able to provide both rapid *and* reliable convergence.

The method of Brent combines inverse quadratic interpolation with bisection to provide a powerful method that has been found to be successful on a wide range of difficult problems. The method is easily implemented and a detailed description of the algorithm may be found in Brent (1971). Similar algorithms of comparable efficiency have been developed by Dekker (1969).

Experience with Brent's algorithm has shown it to be both reliable and efficient on a wide range of problems. A variation of this method is directly available in the MATLAB toolbox and is called fzero. It may be used as follows:

```
x = fzero('funcname',x0,tol,trace);
```

where funcname is replaced by the name of any system function such as cos, sin, etc., or the name of a function predefined by the user. The initial approximation is x0. The accuracy of the solution is set by tol and if trace is a value greater than 1 a trace of the iterations is given. Only the first two parameters need be given and so an alternative call of this function is given by

```
x=fzero('funcname',x0);
```

To use fzero to solve $(e^x - \cos x)^3 = 0$ with tolerance 0.0005, initial approximation 1.65 and no trace of the iterations we write the MATLAB statement

```
x=fzero('f307',1.65,0.0005);
```

where f307 is defined by

```
function F=f307(x);
F=(exp(x)-cos(x)).^3;
```

The MATLAB script included here allows us first to plot and then solve function f307.

```
x=-4:.1:0.5;
plot(x,f307(x)); grid on;
xlabel('x');ylabel('f(x)');
title('f(x)=(exp(x)-cos(x)) .^3');
root=fzero('f307',1.65,0.00005);
fprintf('The root of this equation is %6.4f\n',root);
```

The output and plot generated by this script is not given. However, the script is provided for reader experimentation.

Before we turn to deal with the problem of finding many roots of a polynomial equation simultaneously, we present a comparative study of the MATLAB toolbox function fzero with the function fnewton. The following functions are considered:

(1) $\sin(1/x) = 0$
(2) $(x - 1)^5 = 0$
(3) $x - \tan x = 0$
(4) $\cos\{(x^2 + 5)/(x^4 + 1)\} = 0$

The results of these comparative studies are given in Table 3.10.1. We see that fnewton is slower and less reliable than fzero.

Table 3.10.1. Time in seconds required to solve equations (1) to (4)
with the same starting point $x = -2$ and accuracy = 0.00005.

Function	1	2	3	4
fnewton	Fail	118.4	Fail	15.9
fzero	26.1	52.6	25.6	13.1

3.11 METHODS FOR FINDING ALL THE ROOTS OF A POLYNOMIAL

The problem of solving polynomial equations is a special one in that these equations contain only combinations of integer powers of x and no other functions. Because of their special structure, algorithms have been developed to find all of the roots of a polynomial equation simultaneously. The function roots is provided in MATLAB. This function sets up the companion matrix for the polynomial and determines its eigenvalues, which can be shown to be the roots of the polynomial. For a description of the companion matrix, see Appendix 1.

In the following sections we describe the methods of Bairstow and Laguerre but do not give a detailed theoretical justification of them. We provide a MATLAB function for Bairstow's method.

3.12 BAIRSTOW'S METHOD

Consider the polynomial

$$a_0x^n + a_1x^{n-1} + a_2x^{n-2} + \ldots + a_n = 0 \tag{3.12.1}$$

Since this is a polynomial equation of degree n it has n roots. A common approach for locating the roots of a polynomial is to find all its quadratic factors. These will have the form

$$x^2 + ux + v \qquad\qquad (3.12.2)$$

where u and v are the constants we wish to determine. Once all the quadratic factors are found it is easy to solve the quadratics to find all the roots of the equation. We now outline the major steps used in Bairstow's method for finding these quadratic factors.

If $R(x)$ is the remainder after the division of polynomial (3.12.1) by the quadratic factor (3.12.2), then there will clearly exist constants $b_0, b_1, b_2, \ldots$ such that the equality (3.12.3) holds

$$(x^2 + ux + v)(b_0 x^{n-2} + b_1 x^{n-3} + b_2 x^{n-4} + \ldots + b_{n-2}) + R(x)$$

$$= x^n + a_1 x^{n-1} + a_2 x^{n-2} + \ldots + a_n \qquad\qquad (3.12.3)$$

where a_0 has been taken as one and $R(x)$ will have the form $rx + s$. To ensure $x^2 + ux + v$ is an exact factor of the polynomial (3.12.1) the remainder $R(x)$ must be zero. For this to be true both r and s must be zero and we must adjust u and v until this is true. Thus since both r and s depend on u and v the problem reduces to solving the equations

$$r(u, v) = 0$$
$$s(u, v) = 0$$

To solve these equations we use an iterative method which assumes some initial approximations u_0 and v_0. Then we require improved approximations u_1 and v_1 where $u_1 = u_0 + \Delta u_0$ and $v_1 = v_0 + \Delta v_0$ such that

$$r(u_1, v_1) = 0$$
$$s(u_1, v_1) = 0$$

or r and s are as close to zero as possible.

Now we wish to find the changes Δu_0 and Δv_0 which will result in this improvement. Consequently we must expand the two equations

$$r(u_0 + \Delta u_0, v_0 + \Delta v_0) = 0$$
$$s(u_0 + \Delta u_0, v_0 + \Delta v_0) = 0$$

using a Taylor's series expansion and neglecting higher powers of Δu_0 and Δv_0. This leads to two approximating linear equations for Δu_0 and Δv_0:

$$r(u_0, v_0) + (\partial r/\partial u)_0 \Delta u_0 + (\partial r/\partial v)_0 \Delta v_0 = 0$$

$$\qquad\qquad (3.12.4)$$

$$s(u_0, v_0) + (\partial s/\partial u)_0 \Delta u_0 + (\partial s/\partial v)_0 \Delta v_0 = 0$$

The subscript 0 denotes that the partial derivatives are calculated at the point u_0, v_0. Once the corrections are found the iteration can be repeated until r and s are sufficiently close to zero. The method we have used here is a two-variable form of Newton's method which will be described in section 3.14 of this chapter.

Clearly this method requires the first-order partial derivatives of r and s with respect to u and v. The form of these is not obvious; however, they may be determined using recurrence relations derived from equating coefficients in (3.12.3) and then differentiating them. The details of this derivation are not given here but a clear description of the process is given by Froberg (1969). Once the quadratic factor is found the same process is applied to the residual polynomial with the coefficients b_i to obtain the remaining quadratic factors. The details of this derivation are not provided here but a MATLAB function bairstow is given below.

```
function [rts,it]=bairstow(a,n,tol)
% a is a vector. Poly is x^n+a(1)*x^(n-1)+a(2)*x^(n-2)+...+a(n)
% tol is accuracy to which the polynomial is satisfied.
% The output is produced as an (n x 2) matrix rts
% Columns 1 & 2 contain the real and imag part of root respectively
it=1;
while n>2
  %Initialise for this loop
  u=1; v=1; st=1;
  while st>tol
    b(1)=a(1)-u; b(2)=a(2)-b(1)*u-v;
    for k=3:n
      b(k)=a(k)-b(k-1)*u-b(k-2)*v;
    end;
    c(1)=b(1)-u; c(2)=b(2)-c(1)*u-v;
    for k=3:n-1
      c(k)=b(k)-c(k-1)*u-c(k-2)*v;
    end;
    %calculate change in u and v
    c1=c(n-1); b1=b(n); cb=c(n-1)*b(n-1);
    c2=c(n-2)*c(n-2); bc=b(n-1)*c(n-2);
    if n>3, c1=c1*c(n-3); b1=b1*c(n-3); end;
    dn=c1-c2;
    du=(b1-bc)/dn; dv=(cb-c(n-2)*b(n))/dn;
    u=u+du; v=v+dv;
    st=norm([du dv]); it=it+1;
  end;
  [r1,r2,im1,im2]=solveq(u,v,n,a);
  rts(n,1:2)=[r1 im1]; rts(n-1,1:2)=[r2 im2];
  n=n-2;
  a(1:n)=b(1:n);
end;
%Solve last quadratic or linear equation
u=a(1); v=a(2);
[r1,r2,im1,im2]=solveq(u,v,n,a);
rts(n,1:2)=[r1 im1];
if n==2
  rts(n-1,1:2)=[r2 im2];
end;
```

This function calls the MATLAB function solveq which is defined as follows:

```
function [r1,r2,im1,im2]=solveq(u,v,n,a);
%this function solves quadratics
if n==1
  r1=-a(1);im1=0;
else
  d=u*u-4*v;
  if d<0
    d=-d;
    im1=sqrt(d)/2; r1=-u/2; r2=r1; im2=-im1;
  elseif d>0
    r1=(-u+sqrt(d))/2; im1=0; r2=(-u-sqrt(d))/2; im2=0;
  else
    r1=-u/2; im1=0; r2=-u/2; im2=0;
  end;
end;
```

We may now use bairstow to solve the specific polynomial equation

$$x^5 - 3x^4 - 10x^3 + 10x^2 + 44x + 48 = 0$$

In this case we take the coefficient vector as c where c = [-3   -10   10   44   48] and if we require accuracy of four decimal places we take tol as 0.00005. The script uses bairstow to solve the given polynomial.

```
c=[-3   -10   10   44   48];
[rts, it]=bairstow(c,5,0.00005);
for i=1:5
  fprintf('\nroot%3.0f Real part=%7.4f',i,rts(i,1));
  fprintf(' Imag part=%7.4f\n',rts(i,2));
end;
```

Note how fprintf is used to provide a clearer output from the matrix rts.

```
root  1 Real part= 4.0000  Imag part= 0.0000

root  2 Real part=-1.0000  Imag part=-1.0000

root  3 Real part=-1.0000  Imag part= 1.0000

root  4 Real part=-2.0000  Imag part= 0.0000

root  5 Real part= 3.0000  Imag part= 0.0000
```

It is interesting to compare Bairstow's method with roots. Table 3.12.1 gives the results of this comparison applied to specific polynomials. The problems p1 to p5 are the polynomials:

p1: $x^5 - 3x^4 - 10x^3 + 10x^2 + 44x + 48 = 0$
p2: $x^3 - 3.001x^2 + 3.002x - 1.001 = 0$
p3: $x^4 - 6x^3 + 11x^2 + 2x - 28 = 0$
p4: $x^7 + 1 = 0$
p5: $x^8 + x^7 + x^6 + x^5 + x^4 + x^3 + x^2 + x + 1 = 0$

The results for these problems are given in Table 3.12.1. Both methods determined the correct roots for all problems although the function **roots** is more efficient.

Table 3.12.1. Time required to obtain all roots (in seconds).

	roots	bairstow
p1:	7	33
p2:	6	19
p3:	6	14
p4	10	103
p5:	11	37

3.13 LAGUERRE'S METHOD

Laguerre's method provides a rapidly convergent procedure for locating the roots of a polynomial. The method was used in the MATLAB function **roots1** but this function has now been removed from MATLAB v4. However, the algorithm is interesting and for this reason it is described in this section.

The method is applied to a polynomial in the form

$$p(x) = x^n + a_1 x^{n-1} + a_2 x^{n-2} + \ldots + a_n$$

Starting with an initial approximation x_1 we apply the iterative formula (3.13.1) to the polynomial $p(x)$

$$x_{i+1} = x_i - np(x_i)/[p'(x_i) \pm \sqrt{\{h(x_i)\}}] \quad \text{for } i = 1, 2, \ldots \qquad (3.13.1)$$

where

$$h(x_i) = (n - 1)[(n - 1)\{p'(x_i)\}^2 - np(x_i)\, p''(x_i)]$$

and n is the degree of the polynomial. The sign taken in (3.13.1) is determined so that it is the same as the sign of $p'(x_i)$.

It is important to give some justification for using a formula with such a complex structure. The reader will notice that if the square root term were not present in (3.13.1) then the iterative form would be similar to that of Newton's method, (3.7.1), and identical to that of Schroder's method, (3.8.1). Thus we would have a method with quadratic convergence for the roots of the polynomial. In fact the more complex structure of (3.13.1) provides third-order convergence and consequently faster convergence than Newton's method. The error is proportional to the cube of the previous error. Thus given an initial approximation the method will converge rapidly to a root of the polynomial which we can denote by r.

To obtain the other roots of the polynomial we divide the polynomial $p(x)$ by the factor $(x - r)$ which provides another polynomial of degree $n - 1$. We can then apply iteration (3.13.1) to this polynomial and repeat the whole procedure again. This is repeated until all roots are found to the required accuracy. The process of dividing by the factor $(x - r)$ is known as deflation and can be performed in a simple and efficient way which is described below.

Since we have a known factor $(x - r)$ then

$$a_0 x^n + a_1 x^{n-1} + a_2 x^{n-2} + ... + a_n$$

$$= (x - r)(b_0 x^{n-1} + b_1 x^{n-2} + b_2 x^{n-3} + ... + b_{n-1}) \qquad (3.13.2)$$

On equating coefficients of the powers of x on both sides we have

$$b_0 = a_0$$
$$b_i = a_i + r b_{i-1} \qquad \text{for } i = 1, 2, ..., n - 1 \qquad (3.13.3)$$

This process is known as synthetic division. Care must be taken here, particularly if the root is found to low accuracy, since ill-conditioning can magnify the effect of small errors in the coefficients of the deflated polynomial.

This completes the description of the method but a few important points should be noted. Assuming sufficient accuracy can be maintained in calculations, the method of Laguerre will converge for any value of the initial approximation. Convergence to complex roots and multiple roots can be achieved but at a slower rate because the convergence rate is linear. In the case of a complex root the value of the function $h(x_i)$ becomes negative and consequently the algorithm must be adjusted to deal with this situation. A key feature that should be considered is that the derivatives of the polynomial can be found efficiently by synthetic division.

Thus to summarise the important features of the algorithm:

(1) The algorithm is third order thus providing rapid convergence to individual roots.
(2) All roots of the polynomial can be found by using synthetic division.
(3) Derivatives can be calculated efficiently using synthetic division.

3.14 SOLVING SYSTEMS OF NON-LINEAR EQUATIONS

The methods considered so far have been concerned with finding one or all the roots of a non-linear algebraic equation with one independent variable. We now consider methods for solving systems of non-linear algebraic equations in which each equation is a function of a specified number of variables. We can write such a system in the form

$$f_i(x_1, x_2, ..., x_n) = 0 \quad \text{for } i = 1, 2, 3, ..., n \tag{3.14.1}$$

A simple method for solving this system of non-linear equations is based on Newton's method for the single equation. To illustrate this procedure we first consider a system of two equations in two variables:

$$f_1(x_1, x_2) = 0$$

$$f_2(x_1, x_2) = 0 \tag{3.14.2}$$

Given initial approximations x_1^0 and x_2^0 for x_1 and x_2, we may find new approximations x_1^1 and x_2^1 as follows:

$$x_1^1 = x_1^0 + \Delta x_1^0$$

$$x_2^1 = x_2^0 + \Delta x_2^0 \tag{3.14.3}$$

These approximations should be such that they drive the values of the functions closer to zero, so that

$$f_1(x_1^1, x_2^1) \approx 0$$

$$f_2(x_1^1, x_2^1) \approx 0$$

or

$$f_1(x_1^0 + \Delta x_1^0, x_2^0 + \Delta x_2^0) \approx 0$$

$$f_2(x_1^0 + \Delta x_1^0, x_2^0 + \Delta x_2^0) \approx 0 \tag{3.14.4}$$

Applying a two-dimensional Taylor's series expansion to (3.14.4) gives

$$f_1(x_1^0, x_2^0) + \{\partial f_1/\partial x_1\}^0 \Delta x_1^0 + \{\partial f_1/\partial x_2\}^0 \Delta x_2^0 + ... \approx 0$$

$$f_2(x_1^0, x_2^0) + \{\partial f_2/\partial x_1\}^0 \Delta x_1^0 + \{\partial f_2/\partial x_2\}^0 \Delta x_2^0 + ... \approx 0 \tag{3.14.5}$$

If we neglect terms involving powers of Δx_1^0 and Δx_2^0 higher than one, then (3.14.5) represents a system of two linear equations in two unknowns. The zero superscript means that the function is to be calculated at the initial approximation and Δx_1^0 and Δx_2^0 are the unknowns we wish to find. Having solved (3.14.5) we can obtain our new improved approximations and then repeat the process until we have obtained the accuracy we require. A common convergence criterion is to continue iterations until

$$\sqrt{\{(\Delta x_1^r)^2 + (\Delta x_2^r)^2\}} < \varepsilon$$

where r denotes the iteration number and ε is a small positive quantity preset by the user.

It is a simple step to generalise this procedure for any number of variables and equations. We may write the general system of equations as

$$\mathbf{f}(\mathbf{x}) = \mathbf{0}$$

where $\mathbf{f}$ denotes the column vector of n components $(f_1, f_2, ..., f_n)^T$ and $\mathbf{x}$ is a column vector of n components $(x_1, x_2, ..., x_n)^T$. Let $\mathbf{x}^{r+1}$ denote the value of $\mathbf{x}$ at the $(r + 1)$th iteration; then

$$\mathbf{x}^{r+1} = \mathbf{x}^r + \Delta\mathbf{x}^r \quad \text{for } r = 0, 1, 2, ...$$

If $\mathbf{x}^{r+1}$ is an improved approximation to $\mathbf{x}$ then

$$\mathbf{f}(\mathbf{x}^{r+1}) \approx \mathbf{0}$$

or

$$\mathbf{f}(\mathbf{x}^r + \Delta\mathbf{x}^r) \approx \mathbf{0} \tag{3.14.6}$$

Expanding (3.14.6) by using an n-dimensional Taylor's series expansion gives

$$\mathbf{f}(\mathbf{x}^r + \Delta\mathbf{x}^r) = \mathbf{f}(\mathbf{x}^r) + \nabla\mathbf{f}(\mathbf{x}^r)\,\Delta\mathbf{x}^r + ... \tag{3.14.7}$$

where ∇ is a vector operator of partial derivatives, with respect to each of the n components of $\mathbf{x}$. If we neglect higher-order terms in $(\Delta\mathbf{x}^r)^2$ this gives, by virtue of (3.14.6),

$$\mathbf{f}(\mathbf{x}^r) + \mathbf{J}_r\,\Delta\mathbf{x}^r \approx \mathbf{0} \tag{3.14.8}$$

where $\mathbf{J}_r = \mathbf{f}(\mathbf{x}^r)$. $\mathbf{J}_r$ is called the Jacobian matrix. The subscript r denotes that the matrix is evaluated at the point $\mathbf{x}^r$ and it can be written in component form as

$$\mathbf{J}_r = [\partial f_i(\mathbf{x}^r)/\partial x_j] \quad \text{for } i = 1, 2, ..., n \text{ and } j = 1, 2, ..., n$$

On solving (3.14.8) we have the improved approximation

$$\mathbf{x}^{r+1} = \mathbf{x}^r - \mathbf{J}_r^{-1}\,\mathbf{f}(\mathbf{x}^r) \quad \text{for } r = 0, 1, 2, ...$$

The matrix $\mathbf{J}_r$ may be singular and in this situation the inverse, $\mathbf{J}_r^{-1}$, cannot be calculated.

This is the general form of Newton's method. However, there are two major disadvantages with this method:

(1) The method may not converge unless the initial approximation is a good one.

(2) The method requires the user to provide the derivatives of each function with respect to each variable. The user must therefore provide n^2 derivatives and any computer implementation must evaluate the n functions and the n^2 derivatives at each iteration.

The MATLAB function newtonmv given here implements this method.

```
function [xv,it]=newtonmv(x,f,jf,n,tol)
% User must supply initial approximations for the n variables x,
% the definitions of the functions of the system of equations
% and the partial derivatives in the form of the Jacobian matrix.
it=0; xv=x;
fr=feval(f,xv);
while norm(fr)>tol
  Jr=feval(jf,xv);
  xv1=xv-Jr\fr; xv =xv1;
  fr=feval(f,xv);
  it=it+1;
end;
```

Fig. 3.14.1 illustrates the following system of two equations in two variables:

$$x^2 + y^2 = 4$$

$$xy = 1$$

(3.14.9)

To solve the system (3.14.9) we define it by MATLAB function f308 and its Jacobian by f309 as follows:

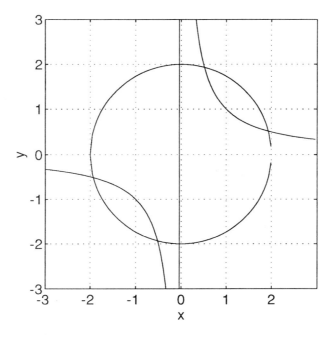

Fig. 3.14.1. Plot of system (3.14.9). Intersections show roots.

```
function f=f308(v)
x=v(1);y=v(2);
% set f vector to size required by function newtonmv.
f=zeros(2,1);
f(1)=x^2+y^2-4;
f(2)=x*y-1;

function jf=f309(v)
x=v(1);y=v(2);
% set jf matrix to size required by function newtonmv.
jf=zeros(2,2);
% each row of jf is assigned the appropriate partial derivatives.
jf(1,:)=[2*x 2*y];
jf(2,:)=[y x];
```

The call of newtonmv to solve the system (3.14.9) using initial approximations for the roots $x = 3$, $y = -1.5$ and a tolerance of 0.00005 is

```
»[rootvals,iter]=newtonmv([3 -1.5]','f308','f309',2,0.00005)
```

This call results in the MATLAB output

```
rootvals =
    1.9319
    0.5176

iter =
    5
```

The solution is $x = 1.9319$ and $y = 0.5176$. Clearly the user must supply a large amount of information for this function and the next section attempts to deal with this problem.

3.15 BROYDEN'S METHOD FOR SOLVING NON-LINEAR EQUATIONS

The method of Newton described in section 3.14 does not provide a practical procedure for solving any but the smallest systems of non-linear equations. As we have seen, the method requires the user to provide not only the function definitions but also the definitions of the n^2 partial derivatives of the functions. Thus, for a system of 10 equations in 10 unknowns, the user must provide 110 function definitions!

To deal with this problem a number of techniques have been proposed but the group of methods which appears most successful is the class of methods known as the quasi-Newton methods. The quasi-Newton methods avoid the calculation of the partial derivatives by obtaining approximations to them involving only the function values. The set of derivatives of the functions evaluated at any point $\mathbf{x}^r$ may be written in the form of the Jacobian matrix

$$\mathbf{J}_r = [\partial f_i(\mathbf{x}^r)/\partial x_j] \quad \text{for } i = 1, 2, ..., n \text{ and } j = 1, 2, ..., n \qquad (3.15.1)$$

The quasi-Newton methods provide an updating formula which gives successive approximations to the Jacobian for each iteration. Broyden and others have shown that under specified circumstances these updating formulae provide satisfactory approximations to the inverse Jacobian. The structure of the algorithm suggested by Broyden is:

(1) Input an initial approximation to the solution. Set the counter r to zero.

(2) Calculate or assume an initial approximation to the inverse Jacobian $\mathbf{B}^r$.

(3) Calculate $\mathbf{p}^r = -\mathbf{B}^r \mathbf{f}^r$ where $\mathbf{f}^r = \mathbf{f}(\mathbf{x}^r)$.

(4) Determine the scalar parameter t such that $\|\mathbf{f}(\mathbf{x}^r + t_r\mathbf{p}^r)\| < \|\mathbf{f}^r\|$ where the symbols $\| \ \|$ denote that the norm of the vector is to be taken.

(5) Calculate $\mathbf{x}^{r+1} = \mathbf{x}^r + t_r \mathbf{p}^r$.

(6) Calculate $\mathbf{f}^{r+1} = \mathbf{f}(\mathbf{x}^{r+1})$. If $\|\mathbf{f}^{r+1}\| < \varepsilon$ (where ε is a small preset positive quantity), then exit. If not continue with step (7).

(7) Use the updating formula to obtain the required approximation to the Jacobian

$$\mathbf{B}^{r+1} = \mathbf{B}^r - (\mathbf{B}^r \mathbf{y}^r - \mathbf{p}^r)(\mathbf{p}^r)^\mathrm{T} \mathbf{B}^r /\{(\mathbf{p}^r)^\mathrm{T} \mathbf{B}^r \mathbf{y}^r\} \text{ where } \mathbf{y}^r = \mathbf{f}^{r+1} - \mathbf{f}^r$$

(8) Set $i = i + 1$ and return to step (3).

The initial approximation to the inverse Jacobian $\mathbf{B}$ is usually taken as a scalar multiple of the unit matrix. The success of this algorithm depends on the nature of the functions to be solved and on the closeness of the initial approximation to the solution. In particular step (4) may present major problems. It may be very expensive in computer time and to avoid this t_r is sometimes set as a constant, usually one or smaller. This may reduce the stability of the algorithm but speeds it up.

It should be noted that other updating formulae have been suggested and it is fairly easy to replace the Broyden formula by others in the above algorithm. In general the problem of solving a system of non-linear equations is a very difficult one. There is no algorithm that is guaranteed to work for all systems of equations. For large systems of equations the available algorithms tend to require large amounts of computer time to obtain accurate solutions.

The MATLAB function broyden implements Broyden's method. It should be noted that this avoids the difficulty of implementing step (4) by taking $t_r = 1$.

```
function [xv,it]=broyden(x,f,n,tol)
% Requires function f and an initial approx x for the n variables
fr=zeros(n,1); it=0; xv=x;
%Set initial Br
Br=eye(n);
fr=feval(f, xv);                                    [Script continues ...
```

```
while norm(fr)>tol
  it=it+1;
  pr=-Br*fr;
  tau=1;
  xv1=xv+tau*pr; xv=xv1;
  oldfr=fr; fr=feval(f,xv);
  %Update approximation to Jacobian using Broyden's formula
  y=fr-oldfr; oldBr=Br;
  oyp=oldBr*y-pr; pB=pr'*oldBr;
  for i=1:n
    for j=1:n
      M(i,j)=oyp(i)*pB(j);
    end;
  end;
  Br=oldBr-M./(pr'*oldBr*y);
end;
```

To solve the system (3.14.9) using Broyden's method we call broyden as follows:

```
»[x, iter]= broyden([3,-1.5]','f308',2,0.00005)
```

This results in

```
x =
    0.5176
    1.9319

iter =
    36
```

This is a correct root of system (3.14.9) but it is not the same root as that found by Newton's method, even though the starting values for the iteration were the same.

It is interesting to compare the approximations to the inverse Jacobian produced by this algorithm and the exact inverse, at the last iteration. These are

$$J_{approx}^{-1} = \begin{bmatrix} 0.0724 & -0.5494 \\ -0.2853 & 0.1731 \end{bmatrix} \quad J_{exact}^{-1} = \begin{bmatrix} 0.0747 & -0.5577 \\ -0.2788 & 0.1494 \end{bmatrix}$$

As a second example we consider the following system of equations which are taken from the MATLAB User's Guide (1989):

$$\begin{aligned} \sin x + y^2 + \log_e z &= 7 \\ 3x + 2^y - z^3 &= -1 \\ x + y + z &= 5 \end{aligned} \qquad (3.15.2)$$

The function f310, which implements (3.15.2), is given below:

```
function q=f310(p)
x=p(1); y=p(2); z=p(3);
q=zeros(3,1);
q(1)=sin(x)+y*y+log(z)-7;
q(2)=x*3+2^y-z^3+1;
q(3)=x+y+z-5;
```

The results of solving (3.15.2) are given below. The starting values used were $x = 0$, $y = 2$ and $z = 2$.

```
»x= broyden([0,2,2]','f310',3,0.00005)
x =
      0.5991
      2.3959
      2.0050
```

This shows the method is successful for two problems and does not require the evaluation of the partial derivatives. The reader may be interested in applying the function newtonmv to this problem. Nine first-order partial derivatives will be required.

3.16 COMPARING THE NEWTON AND BROYDEN METHODS

We end our discussion of the solution of non-linear systems of equations by comparing the performance of the functions broyden and newtonmv, developed in sections 3.14 and 3.15 when solving the system (3.14.9). The following script calls both functions, measures the time taken and provides the number of iterations required for convergence.

```
tic;
[x,it]=broyden([3,-1.5]','f308',2,0.00005)
t0=toc;
disp('broyden time =');
disp(t0);
tic;
[x,it]=newtonmv([3,-1.5]','f308','f309',2,0.00005)
t1=toc;
disp('newtonmv time =');
disp(t1);
```

Running this script gives the following output;

```
x =
      0.5176
      1.9319

it =
      36

broyden time =
      6.2333
```

[Output continues...

```
x =
    1.9319
    0.5176

it =
     5

newtonmv time =
    2.0167
```

It should be noted that although a correct solution is found in each case it is a different root.

The first-order partial derivatives are required for the Newton method and this requires a considerable effort on the part of the user. Solving the above problem demonstrates that the relatively simple form of the function broyden is attractive since it relieves the user of this effort. This advantage of the Broyden method must be set against the additional computer time required by it.

In sections 3.14 and 3.15 two relatively simple algorithms are provided for the solution of a very difficult problem. They cannot always be guaranteed to work and for large problems will converge only slowly.

3.17 SUMMARY

The user wishing to solve non-linear equations will find it is an area which can present particular difficulties. It is always possible to devise or meet with problems which particular algorithms either cannot solve or take a long time to solve. For example, it is just not possible for many algorithms to find the roots of the apparently trivial problem $x^{20} = 0$ very accurately. However, the algorithms described, if used with care, provide ways of solving a wide range of problems. MATLAB is well suited for this study because it allows interactive experimentation and graphical insights into the behaviour of methods and functions.

PROBLEMS

3.1. Omar Khyyam (who lived in the 12th century) solved, by geometric means, a cubic equation with the form

$$x^3 - cx^2 + b^2x + a^3 = 0$$

The positive roots of this equation are the x coordinates of points of intersection in the first quadrant of the circle and parabola given below:

$$x^2 + y^2 - (c - a^3/b^2)x + 2by + b^2 - ca^3/b^2 = 0$$

$$xy = a^3/b$$

For $a = 1$, $b = 2$ and $c = 3$ use MATLAB to plot these two functions and note the x coordinates of the points of intersection. By using the MATLAB function fzero solve the cubic equation and hence verify Omar Khyyam's method. *Hint*: You may find it helpful to use the MATLAB function ginput.

3.2. Use the MATLAB function fnewton to find a root of

$$x^{1.4} - \sqrt{x} + 1/x - 100 = 0$$

given an initial approximation 50. Use an accuracy of 1E–4.

3.3. Find the two real roots of $|x^3| + x - 6 = 0$ using MATLAB function fnewton. Use initial approximations –1 and 1 and an accuracy of 1E–4. Plot the function using MATLAB to verify the equation has only two real roots. *Hint:* Take care in finding the derivative of the function.

3.4. Explain why it is relatively difficult to find the root of $\tan x - c = 0$ when c is a large quantity. Use the MATLAB function fnewton, with initial approximations 1.3 and 1.4 and accuracy 1E–4, to find a root of this equation when $c = 5$ and $c = 10$. Compare the number of iterations required in both cases. *Hint*: A MATLAB plot will be useful.

3.5. Find a root of the polynomial $x^5 - 5x^4 + 10x^3 - 10x^2 + 5x - 1 = 0$ correct to four decimal places by using the MATLAB function schroder with $n = 5$ and a starting value $x_0 = 2$. Use MATLAB function fnewton to solve the same problem. Compare the result and the number of iterations using both methods. Use an accuracy of 5E–7.

3.6. Use the simple iterative method to solve the equation $x^{10} = e^x$. Express the equation in the form $x = f(x)$ in different ways and start the iterations with the initial approximation $x = 1$. Compare the efficiency of the formulae you have devised and check your answer(s) using the MATLAB function fnewton.

3.7. The historic Kepler's equation has the form $E - e \sin E = M$. Solve this equation for $e = 0.96727464$, the eccentricity of Halley's comet, and $M = 4.527594E–3$. Use the MATLAB function fnewton, with an accuracy of 0.00005 and a starting value of 1.

3.8. Examine the performance of the MATLAB function fzero for solving $x^{11} = 0$ with an initial value of –1.5 and also 1. Use an accuracy of 1E–5.

3.9. The smallest positive root of the equation

$$1 - x + x^2/(2!)^2 - x^3/(3!)^2 + x^4/(4!)^2 - \dots = 0$$

is 1.4458. By considering in turn only the first four, five and six terms in the series show that a root of the truncated series approaches this result. Use the MATLAB function fzero to derive these results, with an initial value of 1 and an accuracy of 1E–4.

3.10. Reduce the following system of equations to one equation in terms of x and solve the resulting equation using the MATLAB function fnewton.

$$e^{x/10} - y = 0$$
$$2 \log_e y - \cos x = 2$$

Use the MATLAB function newtonmv to solve these equations directly and compare your results. Use an initial approximation $x = 1$ for fnewton and approximations $x = 1$, $y = 1$ for newtonmv and accuracy 1E–4 in both cases.

3.11. Solve the pair of equations below using the MATLAB function broyden, with the starting point $x = 10$, $y = -10$ and accuracy 1E–4.

$$2x = \sin\{(x + y)/2\}$$
$$2y = \cos\{(x - y)/2\}$$

3.12. Solve the two equations given below using the MATLAB functions newtonmv and broyden with the starting point $x = 1$ and $y = 2$ and accuracy 1E–4.

$$x^3 - 3xy^2 = 1/2$$
$$3x^2 y - y^3 = \sqrt{3}/2$$

3.13. The polynomial equation

$$x^4 - (13 + \varepsilon)x^3 + (57 + 8\varepsilon)x^2 - (95 + 17\varepsilon)x + 50 + 10\varepsilon = 0$$

has roots $1, 2, 5, 5 + \varepsilon$. Use the functions bairstow and roots to find all the roots of this polynomial for $\varepsilon = 0.1$, 0.01 and 0.001. What happens as ε becomes smaller? Use an accuracy of 1E–5.

3.14. Employ the MATLAB function bairstow to find all the roots of the following polynomial using an accuracy requirement of 1E–4.

$$x^5 - x^4 - x^3 + x^2 - 2x + 2 = 0$$

3.15. Use the MATLAB function `roots` to find all the roots of the equation

$$t^3 - 0.5 - (\sqrt{3}/2)i = 0 \quad \text{where } i = \sqrt{-1}$$

Compare with the exact solution $\cos\{(\pi/3 + 2\pi k)/3\} + i \sin\{(\pi/3 + 2\pi k)/3\}$ for $k = 0, 1, 2$. Use an accuracy of 1E–4.

3.16. An outline algorithm for the Illinois method for finding a root of $f(x) = 0$ (Dowell and Jarrett, 1971) is as follows:

> For $k = 0, 1, 2, ...$
> $x_{k+1} = x_k - f_k/f[x_{k-1}, x_k]$
> if $f_k f_{k+1} > 0$ set $x_k = x_{k-1}$ and $f_k = \gamma f_{k-1}$
> where $f_k = f(x_k)$, $f[x_{k-1}, x_k] = (f_k - f_{k-1})/(x_k - x_{k-1})$ and $\gamma = 0.5$.

Write a MATLAB function to implement this method. Note that the *regula falsi* method is similar but differs in that γ is taken as one.

3.17. The following iterative formulae can be used to solve the equation $x^2 - a = 0$:

(a) $x_{k+1} = (x_k + a/x_k)/2 \quad k = 0, 1, 2, ...$
(b) $x_{k+1} = (x_k + a/x_k)/2 - (x_k - a/x_k)^2/(8x_k) \quad k = 0, 1, 2, ...$

These iterative formulae are second- and third-order methods for solving this equation. Write a MATLAB script to implement them and compare the number of iterations required to obtain the square root of 100.112 to five decimal places. For the purpose of illustration, use an initial approximation of 1000.

3.18. Show how MATLAB can be used to study chaotic behaviour by considering the iteration

$$x_{k+1} = g(x_k) \quad \text{for } k = 0, 1, 2, ...$$

where

$$g(x) = cx(1 - x)$$

for different values of the constant c. This simple iteration arises from an attempt to solve a simple quadratic equation. However, its behaviour is complex and for some values of c is chaotic. Write a MATLAB script to plot the value of the iterates against the iterate number for this function and study the behaviour of the iterations for $c = 2.8, 3.25, 3.5$ and 3.8. Use an initial value of $x_0 = 0.7$.

3.19. For the functions solved in problems 3.2, 3.3 and 3.7 use the MATLAB function `plotapp`, given in section 3.9, to find approximate solutions for these functions.

4

Differentiation and integration

4.1 INTRODUCTION

Differentiation and integration are the fundamental operations of differential calculus and occur in almost every field of mathematics, science and engineering. Determining the derivative of a function analytically may be tedious but is relatively straight-forward. The inverse of this process, that of determining the integral of a function, can often be difficult analytically or even impossible.

The difficulty of determining the analytical integral for certain functions has encouraged the development of many numerical procedures for determining approximately the value of definite integrals. In many situations the procedures work well because integration is a smoothing process and errors in the approximation tend to cancel each other. However, for certain types of functions, difficulties may arise and these will be examined as part of our discussion of specific numerical methods for the approximate evaluation of definite integrals.

In the next section of this chapter we will show how the derivative of a function may be estimated for a particular value of the independent variable. The numerical approximations for derivatives require only function values. These approximations can be used to great advantage when derivatives are required in a program. Their application saves the program user the task of determining the analytical expressions for these derivatives.

4.2 NUMERICAL DIFFERENTIATION

In this section we present a range of approximations for first- and higher-order derivatives. Before we derive these approximations in detail we give a simple example which illustrates the dangers of the careless or naive use of such derivative approximations. The simplest approximation for the first-order derivative of a given function $f(x)$ arises from the formal definition of the derivative:

$$\frac{df}{dx} = \lim_{h \to 0} \left(\frac{f(x+h) - f(x)}{h} \right) \tag{4.2.1}$$

One interpretation of (4.2.1) is that the derivative of a function $f(x)$ gives the slope of the tangent to the function at the point x.

For small h we obtain the approximation to the derivative:

$$\frac{df}{dx} \approx \left(\frac{f(x+h) - f(x)}{h} \right) \tag{4.2.2}$$

This would appear to imply that the smaller the value of h the better the value of our approximation in (4.2.1). The MATLAB script below plots Fig. 4.2.1 which shows the error for various values of h.

```
x=1; h(1)=.5; hvals=[ ]; dfbydx=[ ];
for i=1:17
  h=h/10; hvals=[hvals h]; dfbydx(i)=(f401(x+h)-f401(x))/h;
end;
exact=9; loglog(hvals,abs(dfbydx-exact),'*');
axis([1e-18 1 1e-8 1e4])
xlabel('h value'); ylabel('error in approximation');
```

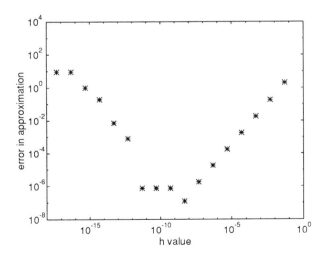

Fig. 4.2.1. A log–log plot showing the error of a simple derivative approximation.

Here the function `f401` is defined by

```
function fv=f401(x)
fv=x.^9;
```

Fig. 4.2.1 shows that for large values of h the error is large but falls rapidly as h is decreased. However, when h becomes less than about 10^{-9} rounding errors dominate and the approximation becomes much worse. Clearly care must be taken in the choice of h. With this warning in mind we develop methods of differing accuracies for any order derivative.

We have seen how a simple approximate formula for the first derivative is easily obtained from the formal definition of the derivative. However, it is difficult to approximate higher derivatives and deduce more accurate formulae in this way and instead we will use the Taylor series expansion of the function $y = f(x)$. To determine the central difference approximation for the derivative of this function at x_i we expand $f(x_i + h)$ thus:

$$f(x_i + h) = f(x_i) + hf'(x_i) + (h^2/2!)f''(x_i)$$

$$+ (h^3/3!)f'''(x_i) + (h^4/4!)f^{(iv)}(x_i) + \dots$$

We sample $f(x)$ at points a distance h apart and write $x_i + h$ as x_{i+1} etc. We will also write $f(x_i)$ as f_i and $f(x_{i+1})$ as f_{i+1}. Thus

$$f_{i+1} = f_i + hf'(x_i) + (h^2/2!)f''(x_i) + (h^3/3!)f'''(x_i)$$

$$+ (h^4/4!)f^{(iv)}(x_i) + \dots \qquad (4.2.3)$$

Similarly

$$f_{i-1} = f_i - hf'(x_i) + (h^2/2!)f''(x_i) - (h^3/3!)f'''(x_i)$$

$$+ (h^4/4!)f^{(iv)}(x_i) - \dots \qquad (4.2.4)$$

We can find an approximation to the first derivative as follows. Subtracting (4.2.4) from (4.2.3) gives

$$f_{i+1} - f_{i-1} = 2hf'(x_i) + 2(h^3/3!)f'''(x_i) + \dots$$

Thus, neglecting terms in h^3 and higher, we have

$$f'(x_i) = (f_{i+1} - f_{i-1})/2h \quad \text{with errors of } O(h^2) \qquad (4.2.5)$$

This is the central difference approximation and differs from (4.2.2) which is a forward difference approximation. Equation (4.2.5) is more accurate than (4.2.2) but in the limit as h approaches zero, they are identical.

To determine an approximation for the second derivative we add (4.2.3) and (4.2.4) to obtain

$$f_{i+1} + f_{i-1} = 2f_i + 2(h^2/2!)f''(x_i) + 2(h^4/4!)f^{(iv)}(x_i) + \dots$$

Thus, neglecting terms in h^4 and higher, we have

$$f''(x_i) = (f_{i+1} - 2f_i + f_{i-1})/h^2 \text{ with errors of } O(h^2) \qquad (4.2.6)$$

By taking more terms in the Taylor series, together with the Taylor series for $f(x + 2h)$ and $f(x - 2h)$ etc., and performing similar manipulations, we can obtain higher derivatives and more accurate approximations if required. Table 4.2.1 gives examples of these formulae.

Table 4.2.1. Derivative approximations.

			Multipliers for $f_{i-3} \dots f_{i+3}$					
	f_{i-3}	f_{i-2}	f_{i-1}	f_i	f_{i+1}	f_{i+2}	f_{i+3}	order of error
$2hf'(x_i)$	0	0	−1	0	1	0	0	h^2
$h^2f''(x_i)$	0	0	1	−2	1	0	0	h^2
$2h^3f'''(x_i)$	0	−1	2	0	−2	1	0	h^2
$h^4f^{(iv)}(x_i)$	0	1	−4	6	−4	1	0	h^2
$12hf'(x_i)$	0	1	−8	0	8	−1	0	h^4
$12h^2f''(x_i)$	0	−1	16	−30	16	−1	0	h^4
$8h^3f'''(x_i)$	1	−8	13	0	−13	8	−1	h^4
$6h^4f^{(iv)}(x_i)$	−1	12	−39	56	−39	12	−1	h^4

The MATLAB function `diffgen` defined below computes the first, second, third and fourth derivative of a given function with errors of $O(h^4)$ for a specified value of x using data from Table 4.2.1.

```
function q = diffgen(func,n,x,h)
if ((n==1)|(n==2)|(n==3)|(n==4))
   c=zeros(4,7);
   c(1,:)=[  0    1  -8     0    8   -1   0];
   c(2,:)=[  0   -1  16   -30   16   -1   0];
   c(3,:)=[1.5 -12  19.5    0 -19.5  12 -1.5];
   c(4,:)=[ -2   24 -78   112  -78   24  -2];
   y=feval(func,x+[-3:3]*h);
   q=c(n,:)*y';  q=q/(12*h^n);
else
   disp('n must be 1, 2, 3 or 4');break
end
```

For example:

```
»result = diffgen('cos',2,1.2,.01)
```

determines the second derivative of $\cos(x)$ for $x = 1.2$ with $h = 0.01$. The following script calls the function diffgen four times to determine the first four derivatives of $y = x^7$ when $x = 1$:

```
h=0.5;i=1;
disp('    h          1st deriv 2nd deriv 3rd deriv  4th deriv');
while h>=1e-5
  t1=h; t2=diffgen('f402', 1, 1, h);
  t3=diffgen('f402', 2, 1, h); t4=diffgen('f402', 3, 1, h);
  t5=diffgen('f402', 4, 1, h);
  fprintf('\n%10.5f%10.5f%10.5f%11.5f%12.5f',t1,t2,t3,t4,t5);
  h=h/10;i=i+1;
end
fprintf('\n')
```

where f402 is given by

```
function fv=f402(x)
fv=x.^7;
```

The output from the above script is

```
»    h          1st deriv 2nd deriv 3rd deriv  4th deriv

  0.50000    1.43750   38.50000   191.62500    840.00000
  0.05000    6.99947   41.99965   209.99816    840.00000
  0.00500    7.00000   42.00000   210.00000    840.00001
  0.00050    7.00000   42.00000   210.00000    839.96425
  0.00005    7.00000   42.00000   209.98692    -59.21189
```

Note that as h is decreased the estimates for the first and second derivatives steadily improve but when $h < 5 \times 10^{-4}$ the estimates for the third and fourth derivatives begin to deteriorate. When $h = 5 \times 10^{-5}$ the estimate for the fourth derivative is totally spurious. In general we cannot predict when this deterioration will begin. The errors are more pronounced using approximations with errors of $O(h^2)$.

4.3 NUMERICAL INTEGRATION

We will begin by examining the definite integral

$$I = \int_a^b f(x)\, dx \qquad (4.3.1)$$

The evaluation of such integrals is often called quadrature and we will develop methods for both finite and infinite values of a and b.

The definite integral (4.3.1) is a summation process but it may also be interpreted as the area under the curve $y = f(x)$ from a to b. Any areas above the x-axis are counted positive; any areas below the x-axis are counted as negative. Many numerical methods for integration are based on using this interpretation to derive approximations to the integral. Typically the interval $[a, b]$ is divided into a number of smaller sub-intervals and by making simple approximations to the curve $y = f(x)$ in the sub-interval the area of the sub-interval may be obtained. The areas of all the sub-intervals are then summed to give an approximation to the integral in the interval $[a, b]$. Variations of this technique are developed by taking groups of sub-intervals and fitting different degree polynomials to approximate $y = f(x)$ in each of these groups. The simplest of these methods is the trapezoidal rule.

The trapezoidal rule is based on the idea of approximating the function $y = f(x)$ in each sub-interval by a straight line so that the shape of the area in the sub-interval is trapezoidal. Clearly, as the number of sub-intervals used increases, the straight lines will approximate the function more closely. Dividing the interval from a to b into n sub-intervals of width h (where $h = (b - a)/n$) we can calculate the area of each sub-interval since the area of a trapezium is its base times the mean of its heights. These heights are f_i and f_{i+1} where $f_i = f(x_i)$. Thus the area of the trapezium is

$$h(f_i + f_{i+1})/2, \text{ for } i = 0, 1, 2, ..., n - 1$$

Summing all the trapezia gives the composite trapezoidal rule for approximating (4.3.1) thus:

$$I \approx h\{(f_0 + f_n)/2 + f_1 + f_2 + ...+ f_{n-1}\}/2 \tag{4.3.2}$$

The truncation error, which is the error due to the implicit approximation in the trapezoidal rule, is

$$E_n \le (b - a)h^2 M/12 \tag{4.3.3}$$

where M is the upper bound for $|f''(t)|$ and t lies between a and b. The MATLAB toolbox function **trapz** implements this procedure and we use it in section 4.4 to compare the performance of the trapezoidal rule with the more accurate Simpson's rule.

The level of accuracy obtained from a numerical integration procedure is dependent on three factors. The first two are the nature of the approximating function and the number of intervals used. These are controlled by the user and give rise to the truncation error, i.e. the error inherent in the approximation. The third factor influencing accuracy is the rounding error, the error caused by the fact that practical computation has limited precision. For a particular approximating function the truncation error will decrease as the number of sub-intervals increases. Integration is a smoothing process and rounding errors do not present a major problem. However, when many intervals are used the time to solve the problem becomes more significant because of the increased amount of computation. This problem may be reduced by writing the script efficiently.

4.4 SIMPSON'S RULE

This rule is based on using a quadratic polynomial approximation to the function $f(x)$ over a pair of sub-intervals, and is illustrated by Fig. 4.4.1. If we integrate the quadratic polynomial passing through the points (x_0, f_0); (x_1, f_1); (x_2, f_2), where $f_1 = f(x_1)$ etc., the following formula is obtained:

$$\int_{x_0}^{x_2} f(x)\, dx = \frac{h}{3}(f_0 + 4f_1 + f_2) \qquad (4.4.1)$$

This is Simpson's rule for one pair of intervals. Applying the rule to all pairs of intervals in the range a to b and adding the results produces the following expression, known as the composite Simpson's rule

$$\int_a^b f(x)\, dx = \frac{h}{3}\{f_0 + 4(f_1 + f_3 + f_5 + \ldots + f_{2n-1}) \\ + 2(f_2 + f_4 + \ldots + f_{2n-2}) + f_{2n}\} \qquad (4.4.2)$$

Here n indicates the number of pairs of intervals and $h = (b - a)/(2n)$. The composite rule may be also be written as a vector product thus:

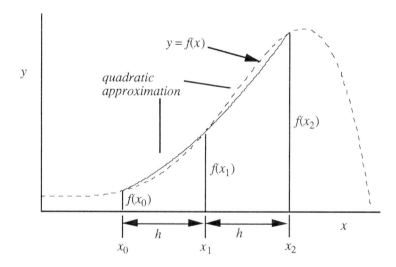

Fig. 4.4.1. Simpson's rule, using a quadratic approximation over two intervals.

$$\int_a^b f(x) \, dx = \frac{h}{3} \mathbf{c} \mathbf{f}^{\mathrm{T}}$$ (4.4.3)

where $\mathbf{c} = [1 \ \ 4 \ \ 2 \ \ 4 \ \ 2 \ ... \ 2 \ \ 4 \ \ 1]$ and $\mathbf{f} = [f_1 \ f_2 \ f_3 \ ... f_{2n}]$.

The error arising from the approximation, called the truncation error, is approximated by

$$E_n \approx (b-a)h^4 f^{(\mathrm{iv})}(t)/180$$

where t lies between a and b. An upper bound for the error is given by

$$E_n \leq (b-a)h^4 M/180$$ (4.4.4)

where M is an upper bound for $|f^{(\mathrm{iv})}(t)|$. The upper bound for the error in the simpler trapezoidal rule, (4.3.3), is proportional to h^2 rather than h^4. This makes Simpson's rule superior to the trapezoidal rule in terms of accuracy at the expense of more function evaluations.

To illustrate different ways of implementing Simpson's rule we provide two alternatives, simp1 and simp2. The function simp1 creates a vector of coefficients v and a vector of function values y and multiplies the two vectors together. Function simp2 provides a more conventional implementation of Simpson's rule. In each case the user must provide the definition of the function to be integrated, the lower and upper limits of integration and the number of sub-intervals to be used. The number of sub-intervals must be an even number since the rule fits a function to a pair of sub-intervals.

```
function q = simp1(func,a,b,n)
if (n/2)~=floor(n/2)
  disp('n must be even');break
end
h=(b-a)/n;
x=[a:h:b]; y=feval(func,x);
v=2*ones(n+1,1);
v2=2*ones(n/2,1);
v(2:2:n)=v(2:2:n)+v2;
v(1)=1; v(n+1)=1;
q=y*v;
q=q*h/3;
```

The second non-vectorised form of this function is

```
function q = simp2(func,a,b,n)
if (n/2)~=floor(n/2)
  disp('n must be even');break
end
```

[*Script continues...*

```
h=(b-a)/n; s=0;
yl=feval(func,a);
for j=2:2:n
  x=a+(j-1)*h; ym=feval(func,x);
  x=a+j*h; yh=feval(func,x);
  s=s+yl+4*ym+yh;
  yl=yh;
end
q=s*h/3;
```

The script shown below calls **trapz** and either **simp1** or **simp2**. These functions can be used to demonstrate the effect on accuracy of the number of pairs of intervals used. The script evaluates the integral of x^7 (defined by **f402** given in section 4.2) in the range 0 to 1.

```
n=2; i=1;
t=clock; flops(0)
disp(' n    integral value');
while n<512
  simpval=simp1('f402',0,1,n); % or simpval=simp2(etc.);
  fprintf('%3.0f%14.9f\n',n,simpval);
  n=2*n;i=i+1;
end
fprintf('\ntime= %4.2f secs  flops=%6.0f\n',etime(clock,t),flops);
```

The output from this script using **simp1** is shown below:

```
» n    integral value
  2    0.171875000
  4    0.129150391
  8    0.125278473
 16    0.125017703
 32    0.125001111
 64    0.125000070
128    0.125000004
256    0.125000000

time= 3.58 secs  flops=  2688
```

On running this script, but using **simp2**, we obtain the same values for the integral but the following results for time and floating point operations:

```
time= 14.78 secs  flops=  2895
```

Equation (4.4.4) shows that the truncation error will decrease rapidly for values of h smaller than 1. The above results illustrate this. The rounding error in Simpson's rule is due to evaluating the function $f(x)$ and the subsequent multiplications and additions. Note also that the vectorised version, **simp1**, is substantially faster than **simp2**, although the number of floating point operations is similar.

We now evaluate the same integral using the MATLAB toolbox function `trapz`. To call this function the user must provide a vector of function values f. The function `trapz(f)` estimates the integral of the function assuming unit spacing between the data points. Thus to determine the integral we multiply `trapz(f)` by the increment h.

```
n=2; i=1;
tic; flops(0)
disp(' n    integral value');
while n<512
   h=1/n; x=0:h:1; f=f402(x);
   trapval=h*trapz(f);
   fprintf('%3.0f%14.9f\n',n,trapval);
   n=2*n; i=i+1;
end
t=toc;
fprintf('\ntime= %4.2f secs   flops=%6.0f\n',t,flops);
```

Running this script gives

```
» n    integral value
  2    0.253906250
  4    0.160339355
  8    0.134043694
 16    0.127274200
 32    0.125569383
 64    0.125142398
128    0.125035603
256    0.125008901

time= 3.85 secs   flops=  2632
```

These results illustrate the fact that the trapezoidal rule is less accurate than Simpson's rule.

4.5 NEWTON–COTES FORMULAE

Simpson's rule is an example of a Newton–Cotes formula for integration. Other examples of these formulae can be obtained by fitting higher degree polynomials through the appropriate number of points. In general we fit a polynomial of degree n through $n + 1$ points. The resulting polynomial can then be integrated to provide an integration formula. Here are some examples of Newton–Cotes formulae together with estimates of their truncation errors.

For $n = 3$ we have

$$\int_{x_0}^{x_3} f(x)\,\mathrm{d}x = \frac{3h}{8}(f_0 + 3f_1 + 3f_2 + f_3) + \text{truncation error } \frac{3h^5}{80}f^{(\mathrm{iv})}(t) \qquad (4.5.1)$$

where t lies in the interval x_0 to x_3.

For $n = 4$ we have

$$\int_{x_0}^{x_4} f(x)\,\mathrm{d}x = \frac{2h}{45}(7f_0 + 32f_1 + 12f_2 + 32f_3 + 7f_4)$$

$$+ \text{truncation error } \frac{8h^7}{945}f^{(\mathrm{vi})}(t)$$

(4.5.2)

where t lies in the interval x_0 to x_4. Composite rules can be generated for both rules (4.5.1) and (4.5.2).

The truncation errors indicate that some improvement in accuracy may be obtained by using these rules rather than Simpson's rule. However, the rules are more complex; consequently greater computational effort is involved and rounding errors may become a more significant problem. The MATLAB toolbox function quad uses an adaptive recursive Simpson's rule and the function quad8 uses an adaptive recursive Newton–Cotes eight panel rule.

The performance of quad and quad8 is compared when evaluating the integral of e^x from 0 to 10 using the following script:

```
t=clock;
flops(0)
quadval=quad('exp',0,10); % or quadval=quad8(etc.); etc.
fprintf('Value of integral %14.9f\n',quadval);
fprintf('\ntime= %4.2f secs  flops=%6.0f\n',etime(clock,t),flops);
```

Running the script calling the function quad('exp',0,10) gives

```
»Value of integral 22025.538517818
```

```
time= 7.00 secs  flops=    856
```

Running the script calling the function quad8('exp',0,10) gives

```
»Value of integral 22025.465894575
```

```
time= 3.78 secs  flops=    328
```

Running the script calling the function simp1('exp',0,10,256) gives

```
»Value of integral 22025.466079655
```

```
time= 0.43 secs  flops=   1311
```

The exact value of the integral is $e^{10} - 1 = 22025.465794807$. Clearly quad8 is faster and more accurate than quad. Using 256 intervals, simp1 is less accurate than quad8 but faster.

4.6 ROMBERG INTEGRATION

A major problem which arises with the non-adaptive Simpson's or Newton–Cote's rules is that the number of intervals required to provide the required accuracy is initially unknown. Clearly one approach to this problem is to double successively the number of intervals used and compare the results of applying a particular rule, as illustrated by the examples in section 4.4. Romberg's method provides an organised approach to this problem and utilises the results obtained by applying Simpson's rule with different interval sizes to reduce the truncation error.

Romberg integration may be formulated as follows. Let I be the exact value of the integral and T_i the approximate value of the integral obtained using Simpson's rule with i intervals. Consequently we may write an approximation for the integral I which includes contributions from the truncation error as follows (note that the error terms are expressed in powers of h^4):

$$I = T_i + c_1 h^4 + c_2 h^8 + c_3 h^{12} + ... \tag{4.6.1}$$

If we double the number of intervals h is halved, giving

$$I = T_{2i} + c_1 (h/2)^4 + c_2 (h/2)^8 + c_3 (h/2)^{12} + ... \tag{4.6.2}$$

We can eliminate the terms in h^4 by subtracting (4.6.1) from 16 times (4.6.2), giving

$$I = (16T_{2i} - T_i)/15 + k_2 h^8 + k_3 h^{12} + ... \tag{4.6.3}$$

Notice that the dominant or most significant term in the truncation error is now of order h^8. In general this will provide a significantly improved approximation to I. For the remainder of this discussion it is advantageous to use a double subscript notation. If we generate an initial set of approximations by successively halving the interval we may represent them by $T_{0,k}$ where $k = 0, 1, 2, 3, 4, ...$. These results may be combined in a similar manner to that described in (4.6.3) by using the general formula

$$T_{r,k} = (16^r T_{r-1,k+1} - T_{r-1,k})/(16^r - 1) \tag{4.6.4}$$

for $k = 0, 1, 2, 3, ...$ and $r = 1, 2, 3, ...$

Here r represents the current set of approximations we are generating. The calculations may be tabulated as follows:

$T_{0,0}$ $T_{0,1}$ $T_{0,2}$ $T_{0,3}$ $T_{0,4}$

$T_{1,0}$ $T_{1,1}$ $T_{1,2}$ $T_{1,3}$

$T_{2,0}$ $T_{2,1}$ $T_{2,2}$

$T_{3,0}$ $T_{3,1}$

$T_{4,0}$

In this case the interval has been halved four times to generate the first five values in the table denoted by $T_{0,k}$. The formula for $T_{r,k}$ given above is used to calculate the remaining values in the table and at each stage the order of the truncation error is increased by four. A common alternative is to write the above table with the rows and columns interchanged.

At each stage the interval size is given by

$$h = (b - a)/2^k \text{ for } k = 0, 1, 2, \ldots \qquad (4.6.5)$$

Romberg integration is implemented in the MATLAB function romb, listed below:

```
function W=romb(func,a,b,d)
T=zeros(d+1,d+1);
for k=1:d+1
  n=2^k; T(1,k)=simp1(func,a,b,n);
end
for p=1:d
  q=4^(p+1);
  for k=0:d-p
    T(p+1,k+1)=(q*T(p,k+2)-T(p,k+1))/(q-1);
  end
end
% for i=1:d+1
%    table=T(i,1:d-i+2); disp(table)
% end
W=T(d+1,1);
```

It is sometimes instructive to output the table of values. This can be done by removing the comment symbol, %, from the three lines in the above script.

We will now apply function romb to the evaluation of $x^{0.1}$ (defined by f403) in the range 0 to 1. The function f403 is given by

```
function fv=f403(x)
fv=x.^0.1;
```

The call of the function romb is

```
»int=romb('f403',0,1,5);
```

Calling this function with the comment symbols from romb removed gives the following output. Note that the best estimate is in the last row of the table.

```
ans =
    0.7887    0.8529    0.8829    0.8969    0.9034    0.9064
    0.8572    0.8849    0.8978    0.9038    0.9066
    0.8853    0.8980    0.9039    0.9067
    0.8980    0.9039    0.9067
    0.9039    0.9067
    0.9067
```

This integral is a surprisingly difficult one and obtaining an accurate result presents a significant problem. The exact solution to four decimal places is 0.9090 so the Romberg method gives only two places of accuracy. However, taking $n = 10$ does give the answer correct to four places thus:

```
»romb('f403',0,1,10)
ans =
    0.9090
```

Generally the Romberg method is very efficient and accurate. For example, it evaluates the integral of e^x from 0 to 10 using five divisions of the interval more accurately and slightly more quickly than the function quad8 with the default tolerance.

An interesting exercise for the reader is to convert function romb to work with the MATLAB function trapz instead of simp1.

4.7 GAUSSIAN INTEGRATION

The common feature of the methods considered so far is that the integrand is evaluated at equal intervals within the range of integration. In contrast, Gaussian integration requires the evaluation of the integrand at specified, but unequal, intervals. For this reason Gaussian integration cannot be applied to data values that are sampled at equal intervals of the independent variable. The general form of the rule is

$$\int_{-1}^{1} f(x)\, dx = \sum_{i=1}^{n} A_i f(x_i) \tag{4.7.1}$$

The parameters A_i and x_i are chosen so that, for a given n, the rule is exact for polynomials up to and including degree $2n - 1$. It should be noticed that the range of integration is required to be from -1 to 1. This does not restrict the integrals to which Gaussian integration can be applied since if $f(x)$ is to be integrated in the range a to b then it can be replaced by the function $g(t)$ integrated from -1 to 1 where

$$t = (2x - a - b)/(b - a)$$

Note that in the above formula, when $x = a$, $t = -1$ and when $x = b$, $t = 1$.

We will now determine the four parameters A_i and x_i for $n = 2$ in (4.7.1). Thus (4.7.1) now becomes

$$\int_{-1}^{1} f(x)\, dx = A_1 f(x_1) + A_2 f(x_2) \tag{4.7.2}$$

This integration rule will be exact for polynomials up to and including degree 3 by ensuring the rule is exact for the polynomials $1, x, x^2$ and x^3 in turn. Thus four equations are obtained as follows:

$$f(x) = 1 \quad \text{gives} \quad \int_{-1}^{1} 1 \, dx = 2 = A_1 + A_2$$

$$f(x) = x \quad \text{gives} \quad \int_{-1}^{1} x \, dx = 0 = A_1 x_1 + A_2 x_2$$

$$f(x) = x^2 \quad \text{gives} \quad \int_{-1}^{1} x^2 \, dx = \frac{2}{3} = A_1 x_1^2 + A_2 x_2^2$$

$$f(x) = x^3 \quad \text{gives} \quad \int_{-1}^{1} x^3 \, dx = 0 = A_1 x_1^3 + A_2 x_2^3$$

Solving these equations gives

$$x_1 = -1/\sqrt{3} \qquad x_2 = 1/\sqrt{3} \qquad A_1 = 1 \qquad A_2 = 1$$

Thus

$$\int_{-1}^{1} f(x) \, dx = f\left(-\frac{1}{\sqrt{3}}\right) + f\left(\frac{1}{\sqrt{3}}\right) \tag{4.7.3}$$

Notice that this rule, like Simpson's rule, is exact for cubics but it requires fewer function evaluations.

A general procedure for obtaining the values of A_i and x_i is based on the fact that in the range of integration it can be shown that $x_1, x_2, ..., x_n$ are the roots of the Legendre polynomial of degree n. The values of A_i can then be obtained from an expression involving the Legendre polynomial of degree n, evaluated at x_i. Tables have been produced for the values of x_i and A_i for various values of n, see Abramowitz and Stegun (1965). Abramowitz and Stegun provide an excellent reference not only for these functions but for a very extensive range of mathematical functions.

Function fgauss defined below performs Gaussian integration. It includes a substitution so that integration in the range a to b is converted to an integration in the range -1 to 1.

```
function q=fgauss(func,a,b,n)
if ((n==2)|(n==4)|(n==8)|(n==16))
  c=zeros(8,4); t=zeros(8,4);
  c(1,1)=1;
  c(1:2,2)=[.6521451548; .3478548451];
  c(1:4,3)=[.3626837833; .3137066458; .2223810344; .1012285362];
  c(:,4)= [.1894506104; .1826034150; .1691565193; .1495959888; ...
           .1246289712; .0951585116; .0622535239; .0271524594];
  t(1,1)  = .5773502691;
  t(1:2,2)=[.3399810435; .8611363115];
  t(1:4,3)=[.1834346424; .5255324099; .7966664774; .9602898564];
  t(:,4)=[.0950125098; .2816035507; .4580167776; .6178762444; ...
          .7554044084; .8656312023; .9445750230; .9894009350];
  j=1;
  while j<=4
    if 2^j==n;break;
    else
      j=j+1;
    end
  end
  s=0;
  for k=1:n/2
    x1=(t(k,j)*(b-a)+a+b)/2; x2=(-t(k,j)*(b-a)+a+b)/2;
    y=feval(func,x1)+feval(func,x2);
    s=s+c(k,j)*y;
  end
  q=(b-a)*s/2;
else
  disp('n must be equal to 2, 4, 8 or 16');break
end
```

The script below calls the function fgauss to integrate $x^{0.1}$ from 0 to 1 (defined by f403 given in section 4.6).

```
disp(' n   integral value');
for j=1:4
  n=2^j;
  int=fgauss('f403',0,1,n);
  fprintf('%3.0f%14.9f\n',n,int);
end
```

The output of this script is

```
n    integral value
  2   0.916290737
  4   0.911012914
  8   0.909561226
 16   0.909199952
```

Gaussian integration with $n = 16$ gives a better result than that obtained by Romberg's method with five divisions of the interval.

4.8 INFINITE RANGES OF INTEGRATION

Other formulae of the Gauss type are available to allow us to deal with integrals having a special form and infinite ranges of integration. These are the Gauss–Laguerre and Gauss–Hermite formulae and take the following form.

(a) *Gauss–Laguerre formulae*. This method is developed from (4.8.1) below:

$$\int_0^\infty e^{-x} g(x) \, \mathrm{d}x = \sum_{i=1}^n A_i \, g(x_i) \tag{4.8.1}$$

The parameters A_i and x_i are chosen so that, for a given n, the rule is exact for polynomials up to and including degree $2n - 1$. Considering the case when $n = 2$ we have

$$g(x) = 1 \ \text{ gives} \int_0^\infty e^{-x} \, \mathrm{d}x = 1 = A_1 + A_2$$

$$g(x) = x \ \text{ gives} \int_0^\infty x e^{-x} \, \mathrm{d}x = 1 = A_1 x_1 + A_2 x_2$$

$$\tag{4.8.2}$$

$$g(x) = x^2 \ \text{gives} \int_0^\infty x^2 e^{-x} \, \mathrm{d}x = 2 = A_1 x_1^2 + A_2 x_2^2$$

$$g(x) = x^3 \ \text{gives} \int_0^\infty x^3 e^{-x} \, \mathrm{d}x = 6 = A_1 x_1^3 + A_2 x_2^3$$

Having evaluated the integrals on the left-hand side of equations (4.8.2) we may solve for the four unknowns x_1, x_2, A_1 and A_2 so that (4.8.1) becomes

$$\int_0^\infty e^{-x} g(x) \, \mathrm{d}x = \frac{2 + \sqrt{2}}{4} g(2 - \sqrt{2}) + \frac{2 - \sqrt{2}}{4} g(2 + \sqrt{2})$$

It can be shown that the x_i are the roots of the nth-order Laguerre polynomial and the coefficients A_i can be calculated from an expression involving the derivative of an nth-order Laguerre polynomial evaluated at x_i.

In general we will wish to evaluate integrals of the form

$$\int_0^\infty f(x) \, \mathrm{d}x$$

We may write this integral as

$$\int_0^\infty e^{-x}\{e^x f(x)\}\, dx$$

Thus, using (4.8.1), we have

$$\int_0^\infty f(x)\, dx = \sum_{i=1}^n A_i \exp(x_i) f(x_i) \tag{4.8.3}$$

Equation (4.8.3) allows integrals to be evaluated over an infinite range, assuming that the value of the integral is finite.

The Gauss–Laguerre method is implemented by the MATLAB function galag thus:

```
function s=galag(func,n)
if (n==2)|(n==4)|(n==8)
  c=zeros(8,3); t=zeros(8,3);
  c(1:2,1)=[1.533326033; 4.450957335];
  c(1:4,2)=[.8327391238; 2.048102438; 3.631146305; 6.487145084];
  c(:,3)  =[.4377234105; 1.033869347; 1.669709765; 2.376924702;...
            3.208540913; 4.268575510; 5.818083368; 8.906226215];
  t(1:2,1)=[.5857864376; 3.414213562];
  t(1:4,2)=[.3225476896; 1.745761101; 4.536620297; 9.395070912];
  t(:,3)  =[.1702796323; .9037017768; 2.251086630; 4.266700170;...
            7.045905402; 10.75851601; 15.74067864; 22.86313174];
  j=1;
  while j<=3
    if 2^j==n;break;
    else
       j=j+1;
    end
  end
  s=0;
  for k=1:n
    x=t(k,j); y=feval(func,x);
    s=s+c(k,j)*y;
  end
else
  disp('n must be 2, 4 or 8');break
end
```

Sample values x_i, and the product $A_i \exp(x_i)$, are given in the function definition. A more complete list may be found in Abramowitz and Stegun (1965).

We will now evaluate the integral $\log_e(1 + e^{-x})$ from zero to infinity. This function is defined by the MATLAB function f404 as follows:

```
function fv=f404(x)
fv=log(1+exp(-x));
```

The script below evaluates the integral using the function galag.

```
disp('  n    integral value');
for j=1:3
  n=2^j;
  int=galag('f404',n);
  fprintf('%3.0f%14.9f\n',n,int);
end
```

The output is as follows:

```
 n    integral value
 2    0.822658694
 4    0.822358093
 8    0.822467051
```

Note that the exact result is $\pi^2/12$ (= 0.82246703342411). The eight-point integration formula is accurate to six decimal places!

(b) *Gauss–Hermite formulae*. This method is developed from (4.8.4) below:

$$\int_{-\infty}^{\infty} \exp(-x^2)g(x)\,\mathrm{d}x = \sum_{i=1}^{n} A_i\, g(x_i) \tag{4.8.4}$$

Again the parameters A_i and x_i are chosen so that, for a given n, the rule is exact for polynomials up to and including degree $2n - 1$. Considering the case when $n = 2$ we have:

$$g(x) = 1 \quad \text{gives} \int_{-\infty}^{\infty} \exp(-x^2)\,\mathrm{d}x = \sqrt{\pi} = A_1 + A_2$$

$$g(x) = x \quad \text{gives} \int_{-\infty}^{\infty} x\exp(-x^2)\,\mathrm{d}x = 0 = A_1 x_1 + A_2 x_2$$

$$g(x) = x^2 \text{ gives} \int_{-\infty}^{\infty} x^2\exp(-x^2)\,\mathrm{d}x = \frac{\sqrt{\pi}}{2} = A_1 x_1^2 + A_2 x_2^2 \tag{4.8.5}$$

$$g(x) = x^3 \text{ gives} \int_{-\infty}^{\infty} x^3\exp(-x^2)\,\mathrm{d}x = 0 = A_1 x_1^3 + A_2 x_2^3$$

We have evaluated the integrals on the left-hand side of equations (4.8.5) and may now solve for the four unknowns x_1, x_2, A_1 and A_2 so that (4.8.4) becomes

$$\int_{-\infty}^{\infty} \exp(-x^2)g(x)\,\mathrm{d}x = \frac{\sqrt{\pi}}{2}g\left(-\frac{1}{\sqrt{2}}\right) + \frac{\sqrt{\pi}}{2}g\left(\frac{1}{\sqrt{2}}\right)$$

An alternative approach is to note that x_i are the roots of the nth-order Hermite polynomial $H_n(x)$. The coefficients A_i can then be determined from an expression involving the derivative of the nth-order Hermite polynomial evaluated at x_i.

In general we will wish to evaluate integrals of the form

$$\int_{-\infty}^{\infty} f(x)\, dx$$

We may write this integral as

$$\int_{-\infty}^{\infty} \exp(-x^2)\{\exp(x^2)f(x)\}\, dx$$

and using (4.8.4) we have

$$\int_{-\infty}^{\infty} f(x)\, dx = \sum_{i=1}^{n} A_i \exp(x_i^2)\, f(x_i) \tag{4.8.6}$$

Again care must be taken to apply (4.8.6) only to functions that have a finite integral in the range $-\infty$ to ∞. Extensive tables of x_i and A_i are again given in Abramowitz and Stegun (1965). The MATLAB function gaherm implements Gauss–Hermite integration thus:

```
function s=gaherm(func,n)
if (n==2)|(n==4)|(n==8)|(n==16)
  c=zeros(8,4); t=zeros(8,4);
  c(1,1)  = 1.461141183;
  c(1:2,2)=[1.059964483; 1.240225818];
  c(1:4,3)=[.7645441286; .7928900483; .8667526065; 1.071930144];
  c(:,4)  =[.5473752050; .5524419573; .5632178291; .5812472754; ...
           .6097369583; .6557556729; .7382456223; .9368744929];
  t(1,1)  = .7071067811;
  t(1:2,2)=[.5246476233; 1.650680124];
  t(1:4,3)=[.3811869902; 1.157193712; 1.981656757; 2.930637420];
  t(:,4)  =[.2734810461; .8229514491; 1.380258539; 1.951787991; ...
           2.546202158; 3.176999162; 3.869447905; 4.688738939];
  j=1;
  while j<=4
    if 2^j==n;break;
    else
      j=j+1;
    end
  end
end
```

[Script continues...]

```
s=0;
for k=1:n/2
  x1=t(k,j); x2=-x1;
  y=feval(func,x1)+feval(func,x2);
  s=s+c(k,j)*y;
end
else
  disp('n must be equal to 2, 4, 8 or 16');break
end
```

We will now evaluate the integral

$$\int_{-\infty}^{\infty} \frac{dx}{\left(1+x^2\right)^2}$$

by the Gauss–Hermite method. The integrand is defined by the MATLAB function f405 as follows:

```
function fv=f405(x)
fv=ones(size(x))./(1+x.^2).^2;
```

The script below uses gaherm to integrate this function.

```
disp(' n    integral value');
for j=1:4
  n=2^j;
  int=gaherm('f405',n);
  fprintf('%3.0f%14.9f\n',n,int);
end
```

The results from running this script are

```
n    integral value
  2    1.298792163
  4    1.482336098
  8    1.550273058
 16    1.565939612
```

The exact value of this integral is $\pi/2 = 1.570796...$.

4.9 GAUSS–CHEBYSHEV FORMULAE

We now consider two interesting cases where the sample points x_i and weights w_i are known in a closed or analytical form. The two integrals together with their closed forms are given below:

$$\int_{-1}^{1} \frac{f(x)}{\sqrt{1-x^2}} \, dx = \frac{\pi}{n} \sum_{k=1}^{n} f(x_k) \text{ where } x_k = \cos\left(\frac{(2k-1)\,\pi}{2n}\right)$$

$$\int_{-1}^{1} \sqrt{1-x^2} \, f(x) \, dx = \frac{\pi}{n+1} \sum_{k=1}^{n} \sin^2\left(\frac{k\pi}{n+1}\right) f(x_k) \text{ where } x_k = \cos\left(\frac{k\pi}{n+1}\right)$$

These expressions are members of the Gauss family, in this case Gauss–Chebyshev formulae. Clearly it is extremely easy to use these formulae for integrands of the required form which have a specified $f(x)$. It is simply a matter of evaluating the function at the specified points, multiplying by the appropriate factor and summing these products. A MATLAB script or function can easily be developed and is left as an exercise for the reader. (See problem 4.11.)

4.10 FILON'S SINE AND COSINE FORMULAE

These formulae can be applied to integrals of the form

$$\int_{a}^{b} f(x) \cos kx \, dx \quad \text{and} \quad \int_{a}^{b} f(x) \sin kx \, dx \tag{4.10.1}$$

The formulae are generally more efficient than standard methods for this form of integral. To derive the Filon formulae we first consider an integral of the form

$$\int_{0}^{2\pi} f(x) \cos kx \, dx$$

By the method of undetermined coefficients we can obtain an approximation to this integrand as follows. Let

$$\int_{0}^{2\pi} f(x) \cos x \, dx = A_1 f(0) + A_2 f(\pi) + A_3 f(2\pi) \tag{4.10.2}$$

Requiring that this should be exact for $f(x) = 1$, x and x^2 we have

$$0 = A_1 + A_2 + A_3$$
$$0 = A_2 \pi + A_3 \, 2\pi$$
$$4\pi = A_2 \pi^2 + A_3 \, 4\pi^2$$

Thus $A_1 = 2/\pi$, $A_2 = -4/\pi$ and $A_3 = 2/\pi$. Hence

$$\int_0^{2\pi} f(x) \cos x \, dx = \frac{1}{\pi}[2f(0) - 4f(\pi) + 2f(2\pi)]$$ (4.10.3)

More general results can be developed as follows:

$$\int_0^{2\pi} f(x) \cos kx \, dx = h[A\{f(x_n)\sin kx_n - f(x_0)\sin kx_0\} + BC_e + DC_o]$$

$$\int_0^{2\pi} f(x) \sin kx \, dx = h[A\{f(x_0)\cos kx_0 - f(x_n)\cos kx_n\} + BS_e + DS_o]$$

where $h = (b - a)/n$, $q = kh$ and

$$A = (q^2 + q \sin 2q/2 - 2 \sin^2 q)/q^3$$ (4.10.4)

$$B = 2\{q(1 + \cos^2 q) - \sin 2q\}/q^3$$ (4.10.5)

$$D = 4(\sin q - q \cos q)/q^3$$ (4.10.6)

$$C_o = \sum_{i=1,3,5...}^{n-1} f(x_i)\cos kx_i$$

$$C_e = \frac{1}{2}\{f(x_0)\cos kx_0 + f(x_n)\cos kx_n\} + \sum_{i=2,4,6...}^{n-2} f(x_i)\cos kx_i$$

It can be seen that C_o and C_e are odd and even sums of cosine terms. S_o and S_e are similarly defined with respect to sine terms.

It is important to note that Filon's method, when applied to functions of the form given in (4.10.1), usually gives better results than Simpson's method for the same number of intervals.

Approximations may be used for the expressions for A, B and D given in (4.10.4), (4.10.5) and (4.10.6) by expanding them in series of ascending powers of q. This leads to the following results:

$$A = 2q^2(q/45 - q^3/315 + q^5/4725 - ...)$$

$$B = 2(1/3 + q^2/15 - 2q^4/105 + q^6/567 - ...)$$

$$D = 4/3 - 2q^2/15 + q^4/210 - q^6/11340 + ...$$

When the number of intervals becomes very large, h and hence q become small. As q tends to zero A tends to zero, B tends to 2/3 and D tends to 4/3. Substituting these values into the formula for Filon's method, it can be shown that it becomes equivalent to Simpson's rule. However, in these circumstances the accuracy of Filon's rule may be worse than Simpson's rule owing to the additional complexity of the calculations.

The MATLAB function `filon` implements Filon's method for the evaluation of appropriate integrals. In the parameter list function `func` defines $f(x)$ of (4.10.1) and this is multiplied by $\cos kx$ when `case` = 1 or $\sin kx$ when `case` $\neq$ 1. The parameters 1 and u specify the lower and upper limit of the integral and n specifies the number of divisions required. The script incorporates a modification to the standard Filon method such that the series approximation is used if q is less than 0.1 rather than (4.10.4) to (4.10.6). The justification for this is that as q becomes small, the accuracy of series approximation is sufficient and easier to compute.

```
function int = filon(func,case,k,l,u,n)
if (n/2)~=floor(n/2)
  disp('n must be even');break
else
  h=(u-l)/n;
  q=k*h;q2=q*q;q3=q*q2;
  if q<0.1
    a=2*q2*(q/45-q3/315+q2*q3/4725);
    b=2*(1/3+q2/15+2*q2*q2/105+q3*q3/567);
    d=4/3-2*q2/15+q2*q2/210-q3*q3/11340;
  else
    a=(q2+q*sin(2*q)/2-2*(sin(q))^2)/q3;
    b=2*(q*(1+(cos(q))^2)-sin(2*q))/q3;
    d=4*(sin(q)-q*cos(q))/q3;
  end
  x=[l:h:u]; y=feval(func,x);
  yodd=y(2:2:n); yeven=y(3:2:n-1);
  if case==1
    c=cos(k*x);
    codd=c(2:2:n); co=codd*yodd';
    ceven=c(3:2:n-1);
    ce=(y(1)*c(1)+y(n+1)*c(n+1))/2; ce=ce+ceven*yeven';
    int=h*(a*(y(n+1)*sin(k*u)-y(1)*sin(k*l))+b*ce+d*co);
  else
    s=sin(k*x);
    sodd=s(2:2:n); so=sodd*yodd';
    seven=s(3:2:n-1);
    se=(y(1)*s(1)+y(n+1)*s(n+1))/2; se=se+seven*yeven';
    int=h*(-a*(y(n+1)*cos(k*u)-y(1)*cos(k*l))+b*se+d*so);
  end
end
```

We now test `filon` by integrating $(\sin x)/x$ in the range 1E–10 to 1. The lower limit is set at 1E–10 to avoid the singularity at zero. The function `f406` is defined as follows. Note that `f406` only defines $1/x$, as required by `filon`.

```
function y=f406(x)
y=ones(size(x))./x;
```

The following script uses `filon` and `filonmod` to evaluate the integral. The function `filonmod` removes the ability to switch to the series formula in `filon`.

```
n=4;
disp('  n    Filon no switch  Filon with switch');
while n<=4096
   int1=filonmod('f406',2,1,1e-10,1,n);
   int2=filon('f406',2,1,1e-10,1,n);
   fprintf('%4.0f%17.8e%17.8e\n',n,int1,int2);
   n=2*n;
end
```

Running this script gives

```
» n    Filon no switch   Filon with switch
   4    1.72067549e+06    1.72067549e+06
   8    1.08265940e+05    1.08265940e+05
  16    6.77884667e+03    6.77884667e+03
  32    4.24742208e+02    4.24742207e+02
  64    2.74361110e+01    2.74361124e+01
 128    2.60175423e+00    2.60175321e+00
 256    1.04956252e+00    1.04956313e+00
 512    9.52549009e-01    9.52550585e-01
1024    9.46489412e-01    9.46487290e-01
2048    9.46109716e-01    9.46108334e-01
4096    9.46085291e-01    9.46084649e-01
```

The exact value of the integral is 0.9460831.

In this particular problem, the switch occurs when $n = 16$. The above output shows that the values of the integral obtained with the switch are marginally more accurate. However, it should be noted that experiments carried out by us have shown that for a lower accuracy of computation than that supplied in the MATLAB environment, the accuracy of Filon's method including the switch is significantly better. The user may find it interesting to experiment with the value of q at which the switch occurs. This is currently set at 0.1.

Finally we choose a function which is appropriate for Filon's method and compare the results with Simpson's rule. The function is $\exp(-x/2)\cos(100x)$ integrated between 0 and 2π. In this case the user-defined function required by the function `filon` is `f407` defined below:

```
function y=f407(x)
y=exp(-x./2);
```

```
function y=f407a(x)
y=exp(-x./2).*cos(100*x);
```

Notice that the definition of `f407`, as required by the function `filon`, does not include the $\cos(100x)$ multiplier in the integrand. In contrast the definition of f407a is the complete integrand. The MATLAB script which implements this comparison is

```
n=4;
disp('  n    Simpsons value    Filons value');
while n<=2048
  int1=filon('f407',1,100,0,2*pi,n);
  int2=simp1('f407a',0,2*pi,n);
  fprintf('%4.0f%17.8e%17.8e\n',n,int2,int1);
  n=2*n;
end
```

The results of this comparison are

» n	Simpsons value	Filons value
4	1.91733833e+00	4.55229440e-05
8	-5.73192992e-01	4.72338540e-05
16	2.42801799e-02	4.72338540e-05
32	2.92263624e-02	4.76641931e-05
64	-8.74419731e-03	4.77734109e-05
128	5.55127202e-04	4.78308678e-05
256	-1.30263888e-04	4.78404787e-05
512	4.53408415e-05	4.78381786e-05
1024	4.77161559e-05	4.78381120e-05
2048	4.78309107e-05	4.78381084e-05

The exact value of the integral to 10 significant digits is 4.783810813E–05. In this particular problem the switch to the series approximations does not take place because of the high value of the coefficient k. The output shows that using 2048 intervals, Filon's method is accurate to eight significant digits. In contrast Simpson's rule is accurate to only five significant digits and its behaviour is highly erratic. However, timing the evaluation of this integral shows that Simpson's method is about 25% faster than Filon's method.

4.11 PROBLEMS IN THE EVALUATION OF INTEGRALS

The methods outlined in the previous sections are based on the assumption that the function to be integrated is well behaved. If this is not so then the numerical methods may give poor, or totally useless, results. Problems may occur if:

(1) the function is continuous in the range of integration but its derivatives are discontinuous or singular.
(2) the function is discontinuous in the range of integration.
(3) the function has singularities in the range of integration.
(4) the range of integration is infinite.

It is vital that these conditions are identified because in most cases these problems cannot be dealt with directly by numerical techniques. Consequently some preparation of the integrand is required before the integral can be evaluated by the appropriate numerical method. Case (1) is the least serious condition but since the derivatives of polynomials are continuous, polynomials cannot accurately represent functions with discontinuous derivatives. Ideally the discontinuity or singularity in the derivative should be located and the integral split into a sum of two or more integrals. The procedure is the same in case (2); the position of the discontinuities must be found and the integral split into a sum of two or more integrals, the ranges of which avoid the discontinuities. Case (3) can be dealt with in various ways: using a change of variable, integration by parts and splitting the integral. In case (4) we must use a method suitable for an infinite range of integration (see section 4.8) or by making a substitution.

The following integral, taken from Fox and Mayers (1968), is an example of case (4):

$$I = \int_1^\infty \frac{dx}{x^2 + \cos(x^{-1})} \qquad (4.11.1)$$

This integral can be estimated either by using function $galag$ (using the substitution $y = x - 1$ to give a lower limit of zero) or by substituting $z = 1/x$. Thus $dz = -dx/x^2$ and (4.11.1) may be transformed as follows:

$$I = -\int_1^0 \frac{dz}{1 + z^2\cos(z)} \quad \text{or} \quad I = \int_0^1 \frac{dz}{1 + z^2\cos(z)} \qquad (4.11.2)$$

The integral (4.11.2) can be easily evaluated by any standard method.

We have discussed a number of techniques for numerical integration. It must be said, however, that even the best methods have difficulty with functions which change very rapidly for small changes in the independent variable. An example of this type of function is $\sin(1/x)$. A MATLAB plot of this function is shown in section 3.8. However, this plot does not give a true representation of the function in the range -0.1 to 0.1 because in this range the function is changing very rapidly and the number of plotting points and the screen resolution are inadequate. Indeed, as x tends to zero the frequency of the function tends to infinity. A further difficulty is that the function has a singularity at $x = 0$. If we decrease the range of x then a small section of the function can be plotted and displayed. For example, in the range $x = 2 \times 10^{-4}$ to 2.05×10^{-4} there are approximately 19 cycles of the function $\sin(1/x)$, as shown in Fig 4.11.1, and in this limited range the function can be effectively sampled and plotted. Summarising, the value of this function can change from an extreme positive to an extreme negative value for a relatively small change in x. The consequence of this is that when estimating the integral of the function a great number of divisions of the range of integration are needed to provide the required level of accuracy, particularly for smaller values of x. For this type of problem adaptive integration methods, such as that used by the MATLAB function $quad8$, have been introduced. These methods increase the number of intervals only in those regions where the function is changing very rapidly, thus reducing the overall number of calculations required.

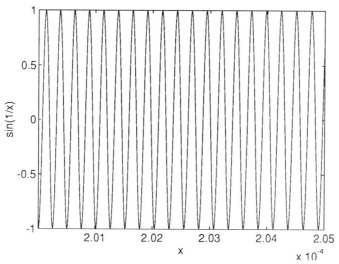

Fig. 4.11.1. Nineteen cycles of the function $\sin(1/x)$
in the range $x = 2 \times 10^{-4}$ to 2.05×10^{-4}.

4.12 TEST INTEGRALS

We now compare the Gauss and Simpson methods of integration with the MATLAB toolbox function quad8 using the following integrals:

$$\int_0^1 x^{0.001}\,dx \;=\; 1000/1001 = 0.999000999... \qquad\qquad (4.12.1)$$

$$\int_0^1 \frac{dx}{1 + (230x - 30)^2} = (\tan^{-1}200 + \tan^{-1}30)/230 = 0.0134924856495 \qquad (4.12.2)$$

$$\int_0^4 x^2(x-1)^2(x-2)^2(x-3)^2(x-4)^2\,dx = 1024/693 = 14.776334776 \qquad (4.12.3)$$

These integrands are defined, respectively, by the MATLAB functions f408, f409, f410 as follows:

```
function y=f408(x)
y=x.^0.001;

function y=f409(x)
y=ones(size(x))./(1+(230*x-30).^2);
```

```
function y=f410(x)
y=(x.^2).*((x-1).^2).*((x-2).^2).*((x-3).^2).*((x-4).^2);
```

To generate the comparative results we define function `ftable` thus:

```
function y=ftable(fname, lowerb,upperb)
tic;
intg=fgauss(fname,lowerb,upperb,16);
tg=toc;
tic;
ints=simp1(fname,lowerb,upperb,2048);
ts=toc;
tic;
intq=quad8(fname,lowerb,upperb,.00005);
tq=toc;
fprintf(fname);
fprintf('%19.8e%15.8e%15.8e\n',intg,ints,intq);
fprintf('time =    %5.2f%16.2f%16.2f\n',tg,ts,tq);
```

The following script applies this function to the three integrals:

```
clear;
disp('function     Gauss          Simpson          quad8');
ftable('f408',0,1);
ftable('f409',0,1);
ftable('f410',0,4);
```

The output from this script is

```
function     Gauss          Simpson          quad8
f408       9.98999164e-01 9.98839883e-01 9.98984176e-01
time =       0.70           10.25            7.72
f409       1.46785766e-02 1.34924856e-02 1.34924856e-02
time =       0.93            1.05           10.25
f410       1.47763051e+01 1.47763348e+01 1.47763348e+01
time =       1.07            1.17            2.68
```

None of integrals (4.12.1) to (4.12.3) are easy to evaluate and Fig. 4.12.1 shows plots of the integrands in the range of integration. It can be seen that each function, at some point, changes rapidly with small changes of the independent variable, making such functions extremely difficult to integrate numerically if a high degree of accuracy is required.

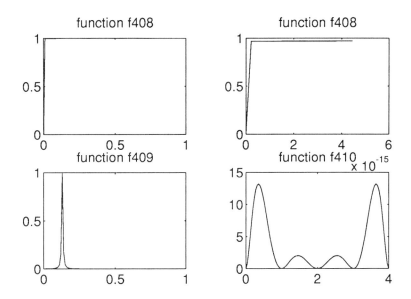

Fig. 4.12.1. Plots of functions f408, f409 and f410, defined in the text,
using the MATLAB function subplot.

4.13 REPEATED INTEGRALS

In this section we confine ourselves to a discussion of repeated integrals using two variables. It is important to note that there is a significant difference between double integrals and repeated integrals. However, it can be shown that if the integrand satisfies certain requirements then double integrals and repeated integrals are equal in value. A detailed discussion of this result is given in Jeffrey (1979).

We have considered in this chapter various techniques for evaluating single integrals. The extension of these methods to repeated integrals can present considerable scripting difficulties. Furthermore the number of computations required for the accurate evaluation of a repeated integral can be enormous. Whilst many algorithms for the evaluation of single integrals can be extended to repeated integrals, here only extensions to the Simpson and Gauss methods with two variables are presented. These have been chosen as the best compromise between programming simplicity and efficiency.

An example of a repeated integral is

$$\int_{a_1}^{b_1} dx \int_{a_2}^{b_2} f(x, y)\, dy$$

(4.13.1)

In this notation the function is integrated with respect to x from a_1 to b_1 and with respect to y from a_2 to b_2. Here the limits of integration are constant but in some applications they may be variables.

4.14 SIMPSON'S RULE FOR REPEATED INTEGRALS

We now apply Simpson's rule to the repeated integral (4.13.1) by applying it first in the y direction and then in the x direction. Consider three equispaced values of y which are y_0, y_1 and y_2. On applying Simpson's rule, (4.4.1), to integration with respect to y in (4.13.1) we have

$$\int_{x_0}^{x_2} dx \int_{y_0}^{y_2} f(x, y)\, dy = \int_{x_0}^{x_2} k\{f(x, y_0) + 4f(x, y_1) + f(x, y_2)\}\, dx \qquad (4.14.1)$$

where $k = y_2 - y_1 = y_1 - y_0$.

Consider now three equispaced values of x: x_0, x_1 and x_2. Applying Simpson's rule again to integration with respect to x, from (4.14.1) we have

$$I \approx hk[f_{0,0} + f_{0,2} + f_{2,0} + f_{2,2} + 4\{f_{0,1} + f_{1,0} + f_{1,2} + f_{2,1}\} + 16f_{1,1}]/9 \qquad (4.14.2)$$

where $h = x_2 - x_1 = x_1 - x_0$ and, for example, $f_{1,2} = f(x_1, y_2)$.

This is Simpson's rule in two variables. By applying this rule to each group of nine points on the surface $f(x, y)$ and summing, the composite Simpson's rule is obtained. The MATLAB function simp2v evaluates repeated integrals in two variables by making direct use of the composite rule.

```
function q=simp2v(func,a,b,c,d,n)
if (n/2)~=floor(n/2)
  disp('n must be even');break
else
  hx=(b-a)/n; x=[a:hx:b];
  hy=(d-c)/n; y=[c:hy:d];
  z=feval(func,x,y);
  v=2*ones(n+1,1); v2=2*ones(n/2,1);
  v(2:2:n)=v(2:2:n)+v2;
  v(1)=1; v(n+1)=1;
  S=v*v'; T=z.*S;
  q=sum(sum(T))*hx*hy/9;
end
```

We will now apply the function simp2v to evaluate the integral

$$\int_0^{10} dx \int_0^{10} \dot{y}^2 \sin x\, dy$$

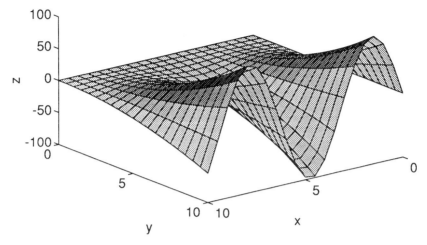

Fig. 4.14.1. Graph of $z = y^2 \sin x$ in the range $x = 0{:}10$, $y = 0{:}10$.

The graph of the function $y^2 \sin x$ is given in Fig. 4.14.1. The function which defines the integrand is as follows:

```
function z=f411(x,y)
[n1,n2]=size(x); [m1,m2]=size(y);
if (n1==1)&(m1==1)
  [xx,yy]=meshgrid(x,y);
  z=yy.^2 .*sin(xx);
else
  disp('x and y must be scalars or row vectors');break
end
```

The required combinations of points is obtained using meshgrid. Using the script

```
disp('   n      integral value     flops/n^2');
n=4; j=1;
while n<=256
  flops(0); int=simp2v('f411',0,10,0,10,n); fl=flops;
  fprintf('%4.0f%17.8e%12.2f\n',n,int,fl/n^2);
  n=2*n; j=j+1;
end
```

gives the following results:

```
  n    integral value    flops/n^2
  4    1.02333856e+03      12.19
  8    6.23187046e+02       9.30
 16    6.13568708e+02       8.07
 32    6.13056704e+02       7.52
 64    6.13025879e+02       7.25
128    6.13023970e+02       7.13
256    6.13023851e+02       7.06
```

The value of this integral exact to four decimal places is 613.0238 and the number of floating point operations tends to $7n^2$. It can be proved, see Salvadori and Baron (1961), that when Simpson's rule is adapted to evaluate repeated integrals the error is still of order h^4 and thus it is possible to use an extrapolation scheme similar to the Romberg method of section 4.6.

4.15 GAUSSIAN INTEGRATION FOR REPEATED INTEGRALS

The Gaussian method can be developed to evaluate repeated integrals with constant limits of integration. In section 4.7 it was shown that for single integrals the integrand must be evaluated at specified points. Thus, if

$$I = \int_{-1}^{1} dx \int_{-1}^{1} f(x, y) \, dy$$

then

$$I \approx \sum_{i=1}^{n} \sum_{j=1}^{m} A_i A_j f(x_i, y_j)$$

The rules for calculating x_i, y_j and A_i are given in section 4.7. The MATLAB function gauss2v evaluates integrals using this technique. Because the values of x and y are chosen on the assumption that the integration takes place in the range -1 to 1 the function includes the necessary manipulations to adjust it so as to accommodate an arbitrary range of integration.

```
function q = gauss2v(func,a,b,c,d,n)
if ((n==2)|(n==4)|(n==8)|(n==16))
  co=zeros(8,4); t=zeros(8,4);
  co(1,1)=1;
  co(1:2,2)=[.6521451548; .3478548451];
  co(1:4,3)=[.3626837833; .3137066458; .2223810344; .1012285362];
  co(:,4)= [.1894506104; .1826034150; .1691565193; .1495959888; ...
            .1246289712;.0951585116; .0622535239; .0271524594];
  t(1,1)   = .5773502691;
  t(1:2,2)=[.3399810435; .8611363115];
  t(1:4,3)=[.1834346424; .5255324099; .7966664774; .9602898564];
  t(:,4)  =[.0950125098; .2816035507; .4580167776; .6178762444; ...
            .7554044084; .8656312023; .9445750230; .9894009350];
```

[Script continues...

```
j=1;
while j<=4
  if 2^j==n;break;
  else
    j=j+1;
  end
end
s=0;
for k=1:n/2
  x1=(t(k,j)*(b-a)+a+b)/2; x2=(-t(k,j)*(b-a)+a+b)/2;
  for p=1:n/2
    y1=(t(p,j)*(d-c)+d+c)/2; y2=(-t(p,j)*(d-c)+d+c)/2;
    z=feval(func,x1,y1)+feval(func,x1,y2)+feval(func,x2,y1);
    z=z+feval(func,x2,y2); s=s+co(k,j)*co(p,j)*z;
  end
end
q=(b-a)*(d-c)*s/4;
else
  disp('n must be equal to 2, 4, 8 or 16');break
end
```

We will now consider the problem of evaluating the following integral:

$$\int_1^2 dx \int_{x^2}^{x^4} x^2 y \, dy \tag{4.15.1}$$

Integrals of this form cannot be estimated directly by the MATLAB function gauss2v, or simp2v, because these functions have been developed for evaluating repeated integrals with constant limits of integration. However, a transformation may be carried out in order to make the limits of integration constant. Let

$$y = (x^4 - x^2)z + x^2 \tag{4.15.2}$$

Thus when $z = 1$, $y = x^4$ and when $z = 0$, $y = x^2$ as required. Differentiating the above expression we have

$$dy = (x^4 - x^2)dz$$

Substituting for y and dy in (4.15.1) we have

$$\int_1^2 dx \int_0^1 x^2\{(x^4 - x^2)z + x^2\}(x^4 - x^2) \, dz \tag{4.15.3}$$

This integral is now in a form that can be integrated using both gauss2v and simp2v. However, we must define a MATLAB function f412 thus:

```
function z=f412(x,y)
[n1,n2]=size(x); [m1,m2]=size(y);
if (n1==1)&(m1==1)
  [xx,yy]=meshgrid(x,y);
  x2=xx.^2;x4=xx.^4; xd=x4-x2;
  z=x2.*(xd.*yy+x2).*xd;
else
  disp('x and y must be scalars or row vectors');break
end
```

This function is used with the functions simp2v and gauss2v in the script below.

```
disp('  n   Simpson value    Gauss value');
n=2; j=1;
while n<=16
  in1 = simp2v('f412',1,2,0,1,n);
  in2 = gauss2v('f412',1,2,0,1,n);
  fprintf('%4.0f%17.8e%17.8e\n',n,in1,in2);
  n=2*n;j=j+1;
end;
```

Running this script gives

n	Simpson value	Gauss value
2	9.54248047e+01	7.65255915e+01
4	8.48837042e+01	8.39728717e+01
8	8.40342951e+01	8.39740259e+01
16	8.39778477e+01	8.39740259e+01

The integral is equal to 83.97402597 (=6466/77). This output shows that in general Gaussian integration is superior to Simpson's rule.

4.16 SUMMARY

In this chapter we have described simple methods for obtaining the approximate derivatives of various orders for specified functions at given values of the independent variable. The results indicate that these methods, although easy to program, are very sensitive to small changes in key parameters and should be used with considerable care. In addition we have given a range of methods for integration. For integration error is not such an unpredictable problem but we must be careful to choose the most efficient method for the integral we wish to evaluate.

PROBLEMS

4.1. Use the function diffgen to find the first and second derivatives of the function $x^2\cos x$ at $x = 1$ using $h = 0.1$ and $h = 0.01$.

4.2. Evaluate the first derivative of $\cos x^6$ for $x = 1, 2$ and 3 using the function `diffgen` and taking $h = 0.001$.

4.3. Write a MATLAB function to differentiate a given function using formulae (4.2.5) and (4.2.6). Use it to solve problems 4.1 and 4.2.

4.4. Find the gradient of $y = \cos x^6$ at $x = 3.1, 3.01, 3.001$ and 3 using the function `diffgen` with $h = 0.001$. Compare your results with the exact result.

4.5. The approximations for partial derivatives may be defined as

$$\partial f / \partial x \approx \{f(x+h,y) - f(x-h,y)\}/(2h)$$

$$\partial f / \partial y \approx \{f(x,y+h) - f(x,y-h)\}/(2h)$$

Write a function to evaluate these derivatives. The function call should have the form

```
[pdx,pdy]=pdiff('func',x,y,h)
```

and determine the partial derivatives of $\exp(x^2 + y^3)$ at $x = 2$, $y = 1$ using this function with $h = 0.005$.

4.6. In a letter sent to Hardy, the Indian mathematician Ramanujan proposed that the number of numbers between a and b which are either squares or sums of two squares is given approximately by the integral

$$0.764 \int_a^b \frac{dx}{\sqrt{\log_e x}}$$

Test this proposition for the following pairs of values of a and b: $(1,10), (1,17)$ and $(1,30)$. You should use the MATLAB function `fgauss` with 16 points to evaluate the integrals required.

4.7. Verify the equality

$$\int_0^\infty \frac{dx}{(1+x^2)(1+r^2x^2)(1+r^4x^2)} = \frac{\pi}{2(1+r+r^3)}$$

for the values of $r = 0, 1, 2$. This result was proposed by Ramanujan. You should use the MATLAB function `galag` for your investigations using eight points.

4.8. Raabe established the result that

$$\int_a^{a+1} \log_e \Gamma(x) \, dx \; = \; a \log_e a - a + \log_e \sqrt{2\pi}$$

Verify this result for $a = 1$ and $a = 2$. Use the MATLAB function simp1 to evaluate the integrals required and the MATLAB function gamma to set up the integrand.

4.9. Use the MATLAB function fgauss with 16 points to evaluate the integral

$$\int_0^1 \frac{\log_e x \, dx}{1 + x^2}$$

Explain why the function fgauss is appropriate for this problem but simp1 is not.

4.10. Use the MATLAB function fgauss with 16 points to evaluate the integral

$$\int_0^1 \frac{\tan^{-1} x}{x} \, dx$$

Note: Integration by parts shows the integrals in problems 4.9 and 4.10 to be the same value except for a sign.

4.11. Write a MATLAB function to implement the formulae given in section 4.9 and use your function to evaluate the following integrals using 10 points for the formula. Compare your results with the Gauss 16 point rule.

(i) $\displaystyle\int_{-1}^1 \frac{e^x}{\sqrt{1-x^2}} \, dx$ (ii) $\displaystyle\int_{-1}^1 e^x \sqrt{1-x^2} \, dx$

4.12. Use the MATLAB function simp1 to evaluate the Fresnel integrals

$$C(1) \; = \; \int_0^1 \cos(\pi t^2) dt \;\; \text{and} \;\; S(1) \; = \; \int_0^1 \sin(\pi t^2) dt$$

Use 32 intervals. The exact values, to seven decimal places, are

$$C(1) = 0.7798934 \text{ and } S(1) = 0.4382591$$

4.13. Use Filon's method, MATLAB function `filon`, with 64 intervals, to evaluate the integral

$$\int_0^\pi \sin x \cos kx \; dx$$

for $k = 0$, 4 and 100. Compare your results with the exact answer $2/(1 - k^2)$.

4.14. Solve problem 4.13 for $k = 100$ using Simpson's rule with 1024 divisions and Romberg's methods with nine divisions.

4.15. Evaluate the following integral using the eight point Gauss–Laguerre method:

$$\int_0^\infty \frac{e^{-x} \; dx}{x + 100}$$

Compare your answer with the exact solution 9.9019419E–3 (103/10402).

4.16. Evaluate the following integral using the 16 point Gauss–Hermite method: Compare your answer with the exact solution $\sqrt{\pi} \exp(-1/4)$.

$$\int_{-\infty}^\infty \exp(-x^2) \cos x \; dx$$

4.17. Evaluate the following integrals, using Simpson's rule for repeated integrals, MATLAB function `simp2v`, using 64 divisions in each direction.

$$\text{(i)} \quad \int_{-1}^1 dy \int_{-\pi}^\pi x^4 y^4 \; dx \quad \text{and} \quad \text{(ii)} \quad \int_{-1}^1 dy \int_{-\pi}^\pi x^{10} y^{10} \; dx$$

4.18. Evaluate the following integrals, using `simp2v`, with 64 divisions in each direction.

$$\text{(i)} \quad \int_0^3 dx \int_1^{\sqrt{x/3}} \exp(y^3) dy \quad \text{and} \quad \text{(ii)} \quad \int_0^2 dx \int_0^{2-x} (1 + x + y)^{-3} \; dy$$

4.19. Evaluate part (ii) of problems 4.17 and 4.18 using Gaussian integration, MATLAB function *gauss2v*. Note: To use this function the range of integration must be constant.

4.20. A definition of the sine integral Si(z) is given by

$$Si(z) = \int_0^z \frac{\sin t}{t} \, dt$$

Evaluate this integral using the 16 point Gauss method for values of $z = 0.5$, 1 and 2. Why does the Gaussian method work and yet the Simpson and Romberg methods fail?

5

Differential equations

5.1 INTRODUCTION

Differential equations arise naturally from attempts to understand how the real world works and from that understanding to predict how it will behave. Essentially, differential equations provide us with a model of some physical situation. This model may be quite simple involving one differential equation or a complex one involving many interrelated differential equations.

To illustrate this situation we will consider a relatively simple problem. Consider the way a hot object cools: for example, a saucepan of milk, the water in a bath or molten iron. Each of these will cool in a different way dependent on the environment but we shall abstract only the most important features that are also easy to model. To model this process by a simple differential equation we use Newton's law of cooling which states that the *rate* at which these objects lose heat as time passes is dependent on the difference between the current temperature of the object and the temperature of its surroundings. This leads to the differential equation

$$dy/dt = K(y - s) \tag{5.1.1}$$

where y is the current temperature at time t, s is the temperature of the surroundings and K is a negative constant for the cooling process.

In addition we require the initial temperature to be specified at time $t = 0$ when the observations begin. Let this be y_0. This fully specifies our model of the cooling process. We only need values for y_0 and K to begin our study. This type of first-order differential equation is called an *initial value problem* because we have an initial value given for the dependent variable y at time $t = 0$. This information, together with values for K and s, is sufficient to specify the problem.

The solution of (5.1.1) is easily obtained analytically and will be a function of t and the constants of the problem. However, there are many differential equations which have no analytic solution or the analytic solution does not provide an explicit relation between y and t. In this situation we use numerical methods to solve the differential equation. This means that we approximate the continuous solution with an approximate discrete solution giving the values of y at specified time steps between the initial value of time and some final time value. Thus we compute values of y, which we denote by y_i, for values of t denoted by t_i where $t_i = t_0 + ih$ for $i = 0, 1, ..., n$. Fig. 5.1.1 illustrates the exact and an approximate solution of (5.1.1) where $K = -0.1$, $s = 10$ and $y_0 = 100$. This figure is generated by using the standard MATLAB function for solving differential equations, ode23, from time 0 to 60 and plotting the steps using the symbol +. The values of the exact solution are plotted on the same graph using the symbol "o".

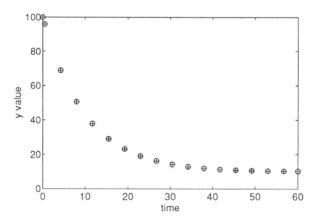

Fig. 5.1.1. Exact "o" and approximate "+" solution for $dy/dt = -0.1(y - 10)$.

To use ode23 to solve (5.1.1) we begin by defining the MATLAB function f501 thus:

```
function yprime=f501(t,y)
yprime=-0.1*(y-10);
```

This function defines the right-hand side of (5.1.1). Then ode23 is called in the following script and requires the initial and final values of t, 0 and 60; a starting value for y of 100; and a tolerance of 0.5.

```
[t y]=ode23('f501',0,60,100,0.5);
plot(t,y,'+');
xlabel('time');ylabel('y value');
hold on;
plot(t,90*exp(-0.1.*t)+10,'o'); % Exact solution.
hold off;
```

This type of step by step solution is based on computing the current y_i value from a single or combination of functions of previous y values. If the value of y is calculated from a combination of more than one previous value it is called a *multi-step* method. If only one previous value is used it is called a *single step* method. We shall now describe a simple *single step* method known as Euler's method.

5.2 EULER'S METHOD

The dependent variable y and the independent variable t, which we have used above, can be replaced by any variable names. For example, many text books use y as the dependent variable and x as the independent variable. However, for some consistency with MATLAB notation we generally use y to represent the dependent variable and t to represent the independent variable. Clearly initial value problems are not restricted to the time domain although, in most practical situations, they are.

Consider the differential equation

$$dy/dt = f(t, y) \tag{5.2.1}$$

One of the simplest approaches for obtaining the numerical solution of a differential equation is the method of Euler. This employs Taylor's series but uses only the first two terms of the expansion. Consider the following form of Taylor's series in which the third term is called the remainder term and represents the contribution of all the terms not included in the series.

$$y(t_0 + h) = y(t_0) + y'(t_0)h + y''(\theta)h^2/2 \tag{5.2.2}$$

where θ lies in the interval (t_0, t_1). For small values of h we may neglect the terms in h^2 and setting $t_1 = t_0 + h$ in (5.2.2) leads to the formula

$$y_1 = y_0 + hy'_0$$

where the prime denotes differentiation with respect to t and $y'_i = y'(t_i)$.

In general

$$y_{n+1} = y_n + hy'_n \text{ for } n = 0, 1, 2, ...$$

By virtue of (5.2.1) this may be written

$$y_{n+1} = y_n + hf(t_n, y_n) \text{ for } n = 0, 1, 2, ... \tag{5.2.3}$$

This is known as Euler's method and it is illustrated geometrically in Fig. 5.2.1. From (5.2.2) we can see that the local truncation error, i.e. the error for individual steps, is of order h^2.

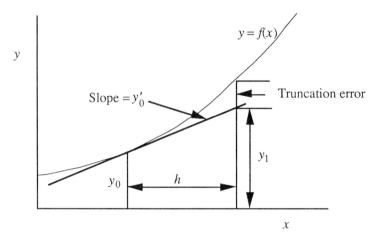

Fig. 5.2.1. Geometric interpretation of Euler's method.

The method is simple to script and a MATLAB function `feuler` is given below.

```
function [tvals, yvals]=feuler(f,start,finish,startval,step)
% Solves dy/dt=f(t,y). start, finish are initial, final values of t
% startval is initial value of y, step is the increment in t
steps=(finish-start)/step+1;
y=startval;t=start;
yvals=startval;tvals=start;
for i=2:steps
  y1=y+step*feval(f,t,y); t1=t+step;
  %collect values together for output
  tvals=[tvals, t1]; yvals=[yvals, y1];
  t=t1;y=y1;
end
```

Applying this function to the differential equation (5.1.1) with $K = 1$, $s = 20$ gives Fig. 5.2.2, which illustrates how the approximate solution varies for different values of h. The exact value computed from the analytical solution is given for comparison purposes by the solid line. Clearly in view of the very large errors shown by Fig. 5.2.2 the Euler method, although simple, requires a very small step h to provide reasonable levels of accuracy. If the differential equation must be solved for a wide range of values of t, the method becomes very expensive in terms of computer time because of the very large number of small steps required to span the interval of interest. In addition, the errors made at each step may accumulate in an unpredictable way. This is a crucial issue and we discuss this in the next section.

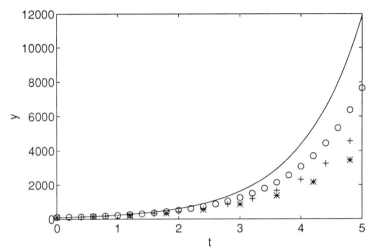

Fig. 5.2.2. Points from the Euler solution $dy/dt = y - 20$ given that
$y = 100$ when $t = 0$. Approximate solutions for $h = 0.2, 0.4$
and 0.6 are plotted using "o", "+" and "*" respectively.
The exact solution is given by the solid line.

5.3 THE PROBLEM OF STABILITY

To ensure that errors do not accumulate we require that the method for solving the differential equation is stable. We have seen that the error at each step in Euler's method is of order h^2. This error is known as the local truncation error since it tells only how accurate the individual step is, not what the error is for a sequence of steps. The error for a sequence of steps is difficult to find since the error from one step affects the accuracy of the next in a way that is often complex. This leads us to the issue of absolute and relative stability. We will discuss these concepts and examine their effects in relation to a simple equation and explain how the results for this equation may be extended to differential equations in general.

Consider the differential equation

$$dy/dt = Ky \qquad (5.3.1)$$

Since $f(t, y) = Ky$, Euler's method will have the form

$$y_{n+1} = y_n + hKy_n \qquad \text{for } n = 0, 1, 2, \dots \qquad (5.3.2)$$

Thus using this recursion repeatedly and assuming that there are no errors in the computation from stage to stage we obtain

$$y_{n+1} = (1 + hK)^{n+1}y_0 \qquad \text{for } n = 0, 1, 2 \dots \qquad (5.3.3)$$

For small enough h it is easily shown that this value will approach the exact value e^{Kt}.

To obtain some understanding of how errors propagate when using Euler's method let us assume that y_0 is perturbed. This perturbed value of y_0 may be denoted by y_0^a where $y_0^a = (y_0 - e_0)$ and e_0 is the error. Thus (5.3.3) becomes, on using this approximate value instead of y_0,

$$y_{n+1}^a = (1 + hK)^{n+1}y_0^a = (1 + hK)^{n+1}(y_0 - e_0) = y_{n+1} - (1 + hK)^{n+1}e_0$$

Consequently the initial error will be magnified if $|1 + hK| \geq 1$. After many steps this initial error will grow and may dominate the solution. This is the characteristic of instability and in these circumstances Euler's method is said to be unstable. If, however, $|1 + hK| < 1$ then the error dies away and the method is said to be absolutely stable. Rewriting this inequality leads to the condition for absolute stability:

$$-2 < hK < 0 \tag{5.3.4}$$

This condition may be too demanding and we may be content if the error does not increase as a proportion of the y values. This is called relative stability. Notice that Euler's method is not absolutely stable for any positive value of K.

The condition for absolute stability can be generalised to an ordinary differential equation of the form of (5.2.1). It can be shown that the condition becomes

$$-2 < h\partial f/\partial y < 0 \tag{5.3.5}$$

This inequality implies that, since $h > 0$, $\partial f/\partial y$ must be negative for absolute stability. Fig. 5.3.1 and Fig. 5.3.2 give a comparison of the absolute and relative error for $h = 0.1$, for the differential equation $dy/dt = y$ where $y = 1$ when $t = 0$. Fig. 5.3.1 shows that the error is increasing rapidly and the errors are large for even relatively small step sizes. Fig. 5.3.2 shows that the error is becoming an increasing proportion of the solution values. Thus the relative error is increasing linearly and so the method is neither relatively stable nor absolutely stable for this problem.

We have seen that Euler's method may be unstable for some values of h. For example, if $K = -100$ then Euler's method is only absolutely stable for $0 < h < 0.02$. Clearly if we required to solve the differential equation between 0 and 10 we would require 500 steps. We now consider an improvement to this method called the trapezoidal method which has improved stability features although it is similar in principle to Euler's method.

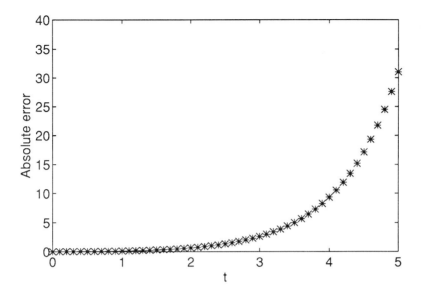

Fig. 5.3.1. Absolute errors in the solution of $dy/dt = y$ where $y = 1$
when $t = 0$, using Euler's method with $h = 0.1$.

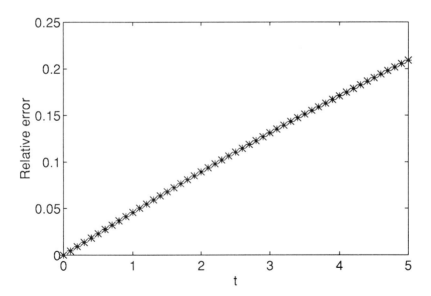

Fig. 5.3.2. Relative errors in the solution of $dy/dt = y$ where $y = 1$
when $t = 0$, using Euler's method with $h = 0.1$.

5.4 THE TRAPEZOIDAL METHOD

The trapezoidal method has the form

$$y_{n+1} = y_n + h\{f(t_n, y_n) + f(t_{n+1}, y_{n+1})\}/2 \quad \text{for } n = 0, 1, 2, \ldots \tag{5.4.1}$$

Applying the error analysis of section 5.3 to this problem gives us, from (5.3.1), that

$$y_{n+1} = y_n + h(Ky_n + Ky_{n+1})/2 \quad \text{for } n = 0, 1, 2, \ldots \tag{5.4.2}$$

Thus expressing y_{n+1} in terms of y_n gives

$$y_{n+1} = \{(1 + hK/2)/(1 - hK/2)\}y_n \quad \text{for } n = 0, 1, 2, \ldots \tag{5.4.3}$$

Using this result recursively for $n = 0, 1, 2, \ldots$ leads to the result

$$y_{n+1} = \{(1 + hK/2)/(1 - hK/2)\}^{n+1} y_0 \tag{5.4.4}$$

Now, as in section 5.3, we can obtain some understanding of how error propagates by assuming that y_0 is perturbed by the error e_0 so that it is replaced by $y_0{}^a = (y_0 - e_0)$. Hence (5.4.4) becomes

$$y_{n+1}{}^a = \{(1 + hK/2)/(1 - hK/2)\}^{n+1}(y_0 - e_0)$$

This leads directly to the result

$$y_{n+1}{}^a = y_{n+1} - \{(1 + hK/2)/(1 - hK/2)\}^{n+1} e_0$$

Thus we conclude from this that the influence of the error term which involves e_0 will die away if its multiplier is less than unity in magnitude, i.e.

$$|(1 + hK/2)/(1 - hK/2)| < 1$$

If K is negative for all h the method is absolutely stable. For positive K it is not absolutely stable for any h.

This completes the error analysis of this method. However, we note that the method requires a value for y_{n+1} before we can start. An estimate for this value can be obtained by using Euler's method, that is

$$y_{n+1} = y_n + hf(t_n, y_n) \quad \text{for } n = 0, 1, 2, \ldots$$

This value can now be used in the right-hand side of (5.4.1) as an estimate for y_{n+1}. This combined method is often known as the Euler–trapezoidal method. The method can be written formally as

(1) Start with n set at zero where n indicates the number of
     steps taken.

(2) Calculate $y^{(1)}{}_{n+1} = y_n + hf(t_n, y_n)$.

(3) Calculate $f(t_{n+1}, y^{(1)}{}_{n+1})$ where $t_{n+1} = t_n + h$.

(5.4.5)

(4) For $k = 1, 2, ...$ calculate

$$y^{(k+1)}{}_{n+1} = y_n + h\{f(t_{n+1}, y^{(k)}{}_{n+1}) + f(t_n, y_n)\}/2$$

At step 4, when the difference between successive values of y_{n+1} is sufficiently small,
increment n by 1 and repeat steps 2, 3 and 4. This method is implemented in MATLAB
function eulertp thus:

```
function [tvals, yvals]=eulertp(f,start,finish,startval,step)
% Solves dy/dt=f(t,y). start, finish are initial, final values of t
% startval is initial value of y, step is the increment in t
steps=(finish-start)/step+1;
y=startval; t=start;
yvals=startval; tvals=start;
for i=2:steps
  y1=y+step*feval(f,t,y);
  t1=t+step;
  loopcount=0; diff=1;
  while abs(diff) >.05
    loopcount=loopcount+1;
    y2=y+step*(feval(f,t,y)+feval(f,t1,y1))/2;
    diff=y1-y2; y1=y2;
  end;
  %collect values together for output
  tvals=[tvals, t1]; yvals=[yvals, y1];
  t=t1; y=y1;
end;
```

We use eulertp to study the performance of this method compared with Euler's
method for solving $dy/dt = y$. The results are given in Fig. 5.4.1 which shows graphs of
the absolute errors of the two methods. The difference is clear but although the Euler–
trapezoidal method gives much greater accuracy for this problem, in other cases the
difference may be less marked. In addition the Euler–trapezoidal method takes longer.

An important feature of this method is the number of iterations that are required to
obtain convergence in step 4. If this is high the method is likely to be inefficient.
However, for the example we have just solved a maximum of two iterations at step 4
was required. This algorithm may be modified to use only one iteration at step 4 in
(5.4.5). This is Heun's method.

Finally we examine theoretically how the error of this method compares with
Euler's method. By considering the Taylor series expansion of y_{n+1} we can obtain the
order of the error in terms of the step size h, thus:

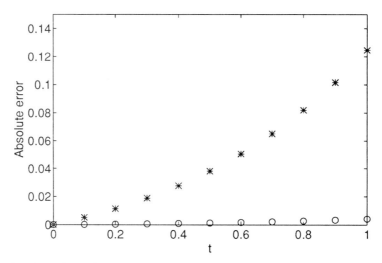

Fig. 5.4.1. Solution of $dy/dt = y$ using Euler "*" and trapezoidal method "o".
Step $h = 0.1$ and $y_0 = 1$ at $t = 0$.

$$y_{n+1} = y_n + hy'_n + h^2y''_n/2! + h^3y'''_n(\theta)/3! \tag{5.4.6}$$

where θ lies in the interval (t_n, t_{n+1}). It can be shown that y''_n may be approximated by

$$y''_n = (y'_{n+1} - y'_n)/h + O(h) \tag{5.4.7}$$

Substituting this expression for y''_n in (5.4.6) gives

$$y_{n+1} = y_n + hy'_n + h(y'_{n+1} - y'_n)/2! + O(h^3)$$

$$= y_n + h(y'_{n+1} + y'_n)/2! + O(h^3)$$

This shows that the local truncation error is of order h^3 so there is a significant improvement in accuracy over the basic Euler method which has a truncation error of order h^2.

We now describe a range of methods which will be considered under the collective title of Runge–Kutta methods.

5.5 RUNGE–KUTTA METHODS

The Runge–Kutta methods comprise a large family of methods having a common structure. Heun's method, described by (5.4.5) but with only one iteration of the corrector, can be recast in the form of a simple Runge–Kutta method. We set

since
$$k_1 = f(t_n, y_n) \text{ and } k_2 = f(t_{n+1}, y_{n+1})$$

We have
$$y_{n+1} = y_n + hf(t_n, y_n)$$

$$k_2 = f(t_{n+1}, y_n + hf(t_n, y_n)) = f(t_{n+1}, y_n + hk_1)$$

Hence from (5.4.1) we have Heun's method in the form

for $n = 0, 1, 2, \ldots$

$$k_1 = f(t_n, y_n)$$
$$k_2 = f(t_{n+1}, y_n + hk_1)$$

and
$$y_{n+1} = y_n + h(k_1 + k_2)/2$$

This is a simple form of a Runge–Kutta method.

The most commonly used Runge–Kutta method is the classical one; it has the form for each step $n = 0, 1, 2, \ldots$

$$k_1 = hf(t_n, y_n)$$
$$k_2 = hf(t_n + h/2, y_n + k_1/2)$$
$$k_3 = hf(t_n + h/2, y_n + k_2/2)$$ (5.5.1)
$$k_4 = hf(t_n + h, y_n + k_3)$$

and
$$y_{n+1} = y_n + (k_1 + 2k_2 + 2k_3 + k_4)/6$$

It has a global error of order h^4.

The next Runge–Kutta method we consider is a variation on the formula (5.5.1). It is due to Gill (1951) and takes the form for each step $n = 0, 1, 2, \ldots$

$$k_1 = hf(t_n, y_n)$$
$$k_2 = hf(t_n + h/2, y_n + k_1/2)$$
$$k_3 = hf(t_n + h/2, y_n + (\sqrt{2} - 1)k_1/2 + (2 - \sqrt{2})k_2/2)$$ (5.5.2)
$$k_4 = hf(t_n + h, y_n - \sqrt{2}k_2/2 + (1 + \sqrt{2}/2)k_3)$$

and
$$y_{n+1} = y_n + \{k_1 + (2 - \sqrt{2})k_2 + (2 + \sqrt{2})k_3 + k_4\}/6$$

Again this method is fourth-order and has a local truncation error of order h^5 and a global error of order h^4.

A number of other forms of the Runge–Kutta method have been derived which have particularly advantageous properties. The equations for these methods will not be given but their important features are as follows:

(1) *Runge–Kutta–Merson method* (Merson, 1957). This method has an error term
     of order h^5 and in addition allows an estimate of the local truncation error to
     be obtained at each step in terms of known values.

(2) *The Ralston–Runge–Kutta method* (Ralston, 1962). We have some degree of
     freedom in assigning the coefficients for a particular Runge–Kutta method.
     In this formula the values of the coefficients are chosen so as to minimise the
     truncation error.

(3) *The Butcher–Runge–Kutta method* (Butcher, 1964). This method provides
     higher accuracy at each step, the error being of order h^6.

Runge–Kutta methods have the general form for each step $n = 0, 1, 2, \ldots$

$$k_1 = hf(t_n, y_n)$$

$$k_i = hf(t_n + hd_i, y_n + \sum_{j=1}^{i-1} c_{ij}k_j) \tag{5.5.3}$$

$$y_{n+1} = y_n + \sum_{j=1}^{p} b_j k_j \tag{5.5.4}$$

The order of this general method is p.

The derivation of the various Runge–Kutta methods is based on the expansion of
both sides of (5.5.4) as a Taylor's series and equating coefficients. This is a relatively
straightforward idea but involves lengthy algebraic manipulation.

We now discuss the stability of the Runge–Kutta methods. Since the instability
which may arise in the Runge–Kutta methods can usually be reduced by a step size
reduction it is known as partial instability. To avoid repeated reduction of the value of
h and rerunning the method, an estimate of the value of h which will provide stability
for the fourth-order Runge–Kutta methods is given by the inequality

$$-2.78 < h\partial f/\partial y < 0$$

In practice $\partial f/\partial y$ may be approximated using the difference of successive values of f and
y.

Finally it is interesting to see how we can apply MATLAB to provide an elegant
function for the general Runge–Kutta method given by (5.5.4) and (5.5.3). We define
two vectors d and b, where d contains the coefficients d_i in (5.5.3) and b contains the
coefficients b_j in (5.5.4), and a matrix c which contains the coefficients c_{ij} in (5.5.3).
If the computed values of the k_j are kept in a vector k then the MATLAB statements which
will generate the values of the function and the new value of y are relatively simple; they
will have the form

```
k(1)=feval('f',t,y);
for i=1:p
  k(i)=step*feval('f',t+step*d(i),y+c(i,1:i-1)*k(1:i-1));
end
y=y+b*k';
```

This is of course repeated for each step. A MATLAB function, rkgen, based on this is
given below. Since c and d are easily changed in the script, any form of Runge–Kutta
method can be implemented using this function and it is useful for experimenting with
different techniques.

```
function[tvals,yvals]=rkgen(f,start,finish,startval,step,method)
% Solves dy/dt=f(t,y). start, finish are initial, final values of t
% startval is initial value of y, step is the increment in t
% method (1, 2 or 3) selects Classical, Butcher or Merson RK.
b=[ ];c=[ ];d=[ ];
if method <1 | method >3
  disp('Method number unknown so using Classical');
  method=1;
end;
if method==1
  order=4;
  b=[ 1/6 1/3 1/3 1/6]; d=[0 .5 .5 1];
  c=[0 0 0 0;0.5 0 0 0;0 .5 0 0;0 0 1 0];
  disp('Classical method selected');
elseif method ==2
  order=6;
  b=[0.07777777778 0 0.355555556 0.13333333 ...
    0.355555556  0.0777777778];
  d=[0 .25 .25 .5 .75 1];
  c(1:4,:)=[0 0 0 0 0 0;0.25 0 0 0 0 0;0.125 0.125 0 0 0 0; ...
    0 -0.5 1 0 0 0];
  c(5,:)=[.1875 0 0 0.5625 0 0];
  c(6,:)=[-.4285714 0.2857143 1.714286 -1.714286 1.1428571 0];
  disp('Butcher method selected');
else
  order=5;
  b=[1/6 0 0 2/3 1/6];
  d=[0 1/3 1/3 1/2 1];
  c=[0 0 0 0 0;1/3 0 0 0 0;1/6 1/6 0 0 0;1/8 0 3/8 0 0; ...
    1/2 0 -3/2 2 0];
  disp('Merson method selected');
end;
steps=(finish-start)/step+1;
y=startval; t=start;
yvals=startval; tvals=start;
```

[Script continues...

```
  for j=2:steps
    k(1)=step*feval(f,t,y);
    for i=2:order
      k(i)=step*feval(f,t+step*d(i),y+c(i,1:i-1)*k(1:i-1)');
    end;
    y1=y+b*k'; t1=t+step;
    %collect values together for output
    tvals=[tvals, t1]; yvals=[yvals, y1];
    t=t1; y=y1;
  end;
```

A further issue that needs to be considered is that of adaptive step size adjustment. Where a function is relatively smooth in the area of interest a large step may be used throughout the region. If the region is such that rapid changes in y occur for small changes in t then a small step size is required. However, for functions were both these regions exist then rather than use a small step in the whole region adaptive step size adjustment would be more efficient. The details of producing this step adjustment are not provided here but for an elegant discussion see Press *et al.* (1990). This type of procedure is implemented for Runge–Kutta methods in the MATLAB toolbox functions ode23 and ode45.

Fig. 5.5.1 plots the relative errors in the solution of the specific differential equation $dy/dt = -y$ by the classical, Merson and Butcher–Runge–Kutta methods using the following MATLAB script:

```
  char=['o'  '*'  '+'];
  for meth=1:3
    [t,x]=rkgen('f502',0,3,1,.25,meth);
    re=(x-exp(-t))./exp(-t);
    plot(t,re,char(meth));
    axis([0 3 0 1.5e-4])
    xlabel('t');ylabel('relative error');
    hold on;
  end;
  hold off;
```

The function f502 is defined thus:

```
  function yprime=f502(t,y)
  yprime=-y;
```

It is clear from the graphs that Butcher's and Merson's methods are significantly more accurate than the classical method and Butcher's method is comfortably the best.

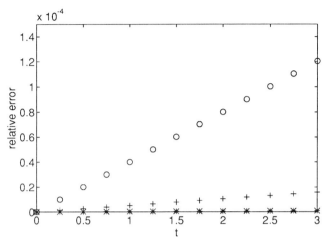

Fig. 5.5.1 Solution of $dy/dt = -y$. The "*" represents Butcher's method,
"+" Merson's method and "o" the classical method.

5.6 PREDICTOR–CORRECTOR METHODS

The trapezoidal method, which has already been described, is a simple example of both a Runge–Kutta method and a predictor–corrector method with a truncation error of order h^3. The predictor–corrector methods we shall consider now have much smaller truncation errors. As an initial example we will consider the Adams–Bashforth–Moulton method. This method is based on the following equations:

$$y_{n+1} = y_n + h(55y'_n - 59y'_{n-1} + 37y'_{n-2} - 9y'_{n-3})/24 \qquad (P)$$
$$y'_{n+1} = f(t_{n+1}, y_{n+1}) \qquad\qquad\qquad\qquad\qquad\qquad\quad (E)$$
$$\qquad\qquad\qquad\qquad\qquad\qquad\qquad\qquad\qquad\qquad\qquad (5.6.1)$$

$$y_{n+1} = y_n + h(9y'_{n+1} + 19y'_n - 5y'_{n-1} + y'_{n-2})/24 \qquad (C)$$
$$y'_{n+1} = f(t_{n+1}, y_{n+1}) \qquad\qquad\qquad\qquad\qquad\qquad\quad (E)$$
$$\qquad\qquad\qquad\qquad\qquad\qquad\qquad\qquad\qquad\qquad\qquad (5.6.2)$$

where $t_{n+1} = t_n + h$. Note that the labels P and E in (5.6.1) denote the predictor equation followed by a function evaluation and the labels C and E in (5.6.2) denote the corrector equation followed by a function evaluation. The truncation error for both the predictor and corrector is $O(h^5)$. The first equation in the system (5.6.1) requires a number of initial values to be known before y can be calculated. After each application of (5.6.1) and (5.6.2), i.e. a complete *PECE* step, the independent variable t_n is incremented by h, n is incremented by one and the process repeated until the differential equation has been solved in the range of interest. The method is started with $n = 3$ and consequently the values of y_3, y_2, y_1 and y_0 must be known before the method can be applied. For this reason it is called a *multi-point method*. In practice y_3, y_2, y_1 and y_0 must be obtained by using a self-starting procedure such as one of the Runge–Kutta methods described in section 5.5. The self-starting procedure chosen should have the same order truncation error as the predictor–corrector method.

The Adams–Bashforth–Moulton method is often used since its stability is relatively good. Its range of absolute stability in *PECE* mode is

$$-1.25 < h\partial f/\partial y < 0$$

Apart from the need for initial starting values the Adams–Bashforth–Moulton method in the *PECE* mode requires less computation at each step than the fourth-order Runge–Kutta method. For a true comparison of these methods, however, it is necessary to consider how they behave over a range of problems since applying any method to some differential equations results, at each step, in a growth of error that ultimately swamps the calculation since the step is outside the range of absolute stability.

The Adams–Bashforth–Moulton method is implemented by the function *abm*. It should be noted that errors arise from the choice of starting procedure, in this case the classical Runge–Kutta method. It is, however, easy to amend this function to include the option of entering highly accurate initial values.

```
function [tvals, yvals]=abm(f,start,finish,startval,step)
%Adams Bashforth Moulton method
%Set up matrices for Runge-Kutta methods
b=[ ];c=[ ];d=[ ]; order=4;
b=[ 1/6 1/3 1/3 1/6]; d=[0 .5 .5 1];
c=[0 0 0 0;0.5 0 0 0;0 .5 0 0;0 0 1 0];
steps=(finish-start)/step+1;
y=startval; t=start; fval(1)=feval(f,t,y);
ys(1)=startval; yvals=startval;tvals=start;
for j=2:4
  k(1)=step*feval(f,t,y);
  for i=2:order
    k(i)=step*feval(f,t+step*d(i),y+c(i,1:i-1)*k(1:i-1)');
  end;
  y1=y+b*k'; ys(j)=y1; t1=t+step;
  fval(j)=feval(f,t1,y1);
  %collect values together for output
  tvals=[tvals,t1]; yvals=[yvals,y1];
  t=t1; y=y1;
end;
%ABM now applied
for i=5:steps
  y1=ys(4)+step*(55*fval(4)-59*fval(3)+37*fval(2)-9*fval(1))/24;
  t1=t+step; fval(5)=feval(f,t1,y1);
  yc=ys(4)+step*(9*fval(5)+19*fval(4)-5*fval(3)+fval(2))/24;
  fval(5)=feval(f,t1,yc);
  fval(1:4)=fval(2:5);
  ys(4)=yc;
  tvals=[tvals,t1]; yvals=[yvals,yc];
  t=t1; y=y1;
end;
```

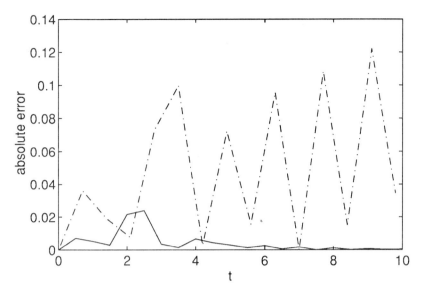

Fig. 5.6.1. Absolute error in solution of $dy/dt = -2y$ using the Adams–Bashforth–Moulton method. The solid line plots the errors with a step size of 0.5. The dot–dashed line plots the errors with step size 0.7.

Fig. 5.6.1 illustrates the behaviour of the Adams–Bashforth–Moulton method when applied to the specific problem $dy/dt = -2y$ where $y = 1$ when $t = 0$, using a step size equal to 0.5 and 0.7 in the interval 0 to 10. It is interesting to note that for this problem, since $\partial f/\partial y = -2$, the range of steps for absolute stability is $0 \leq h \leq 0.625$. For $h = 0.5$, a value inside the range of absolute stability, the plot shows that the absolute error does die away. However, for $h = 0.7$, a value outside the range of absolute stability, the plot shows that the absolute error increases.

5.7 HAMMING'S METHOD AND THE USE OF ERROR ESTIMATES

The method of Hamming (1959) is based on the following pair of predictor–corrector equations:

$$y_{n+1} = y_{n-3} + 4h(2y'_n - y'_{n-1} + 2y'_{n-2})/3 \qquad (P)$$
$$y'_{n+1} = f(t_{n+1}, y_{n+1}) \qquad (E)$$

$$y_{n+1} = \{9y_n - y_{n-2} + 3h(y'_{n+1} + 2y'_n - y'_{n-1})\}/8 \qquad (C)$$
$$y'_{n+1} = f(t_{n+1}, y_{n+1}) \qquad (E)$$

$$(5.7.1)$$

where $t_{n+1} = t_n + h$.

The first equation (P) is used as the predictor and the third as the corrector (C). To obtain a further improvement in accuracy at each step in the predictor and corrector we modify these equations by using expressions for the local truncation errors. Approximations for these local truncation errors can be obtained by using the predicted and corrected values of the current approximation to y. This leads to the equations

$$y_{n+1} = y_{n-3} + 4h(2y'_n - y'_{n-1} + 2y'_{n-2})/3 \qquad (P) \qquad (5.7.2)$$

$$y^M_{n+1} = y_{n+1} - 112(Y_P - Y_C)/121 \qquad (5.7.3)$$

In this equation Y_P and Y_C represent the predicted and corrected value of y at the nth step.

$$y^*_{n+1} = [9y_n - y_{n-2} + 3h\{(y^M)'_{n+1} + 2y'_n - y'_{n-1}\}]/8 \qquad (C) \qquad (5.7.4)$$

In this equation $(y^M)'_{n+1}$ is the value of y'_{n+1} calculated using the modified value of y_{n+1} which is y^M_{n+1}.

$$y_{n+1} = y^*_{n+1} + 9(y_{n+1} - y^*_{n+1})/121 \qquad (5.7.5)$$

Equation (5.7.2) is the predictor and (5.7.3) modifies the predicted value by using an estimate of the truncation error. Equation (5.7.4) is the corrector which is modified by (5.7.5) using an estimate of the truncation error. The equations in this form are each used only once before n is incremented and the steps repeated again. This method is implemented as MATLAB function fhamming thus:

```
function [tvals, yvals]=fhamming(f,start,finish,startval,step)
% Solves dy/dt=f(t,y). start, finish are initial, final values of t
% startval is initial value of y, step is the increment in t
% 3 steps of Runge-Kutta are required so that hamming can start.
% Set up matrices for Runge-Kutta methods
b=[ ];c=[ ];d=[ ];
order=4;
b=[1/6 1/3 1/3 1/6]; d=[0 0.5 0.5 1];
c=[0 0 0 0;0.5 0 0 0;0 0.5 0 0;0 0 1 0];
steps=(finish-start)/step+1;
y=startval;t=start;
fval(1)=feval(f,t,y);
ys(1)=startval;
yvals=startval; tvals=start;
for j=2:4
  k(1)=step*feval(f,t,y);
  for i=2:order
    k(i)=step*feval(f,t+step*d(i),y+c(i,1:i-1)*k(1:i-1)');
  end;
  y1=y+b*k'; ys(j)=y1; t1=t+step; fval(j)=feval(f,t1,y1);
  %collect values together for output
  tvals=[tvals, t1]; yvals=[yvals, y1]; t=t1; y=y1;
end;
```
[Script continues...

```
%Hamming now applied
for i=5:steps
  y1=ys(1)+4*step*(2*fval(4)-fval(3)+2*fval(2))/3;
  t1=t+step; y1m=y1;
  if i>5, y1m=y1+112*(c-p)/121; end;
  fval(5)=feval(f,t1,y1m);
  yc=(9*ys(4)-ys(2)+3*step*(2*fval(4)+fval(5)-fval(3)))/8;
  ycm=yc+9*(y1-yc)/121;
  p=y1; c=yc;
  fval(5)=feval(f,t1,ycm);
  fval(2)=fval(3); fval(3)=fval(4); fval(4)=fval(5);
  ys(1)=ys(2); ys(2)=ys(3); ys(3)=ys(4); ys(4)=ycm;
  tvals=[tvals, t1]; yvals=[yvals, ycm];
  t=t1;
end;
```

The choice of h must be made carefully so that the error does not increase without bound. Fig. 5.7.1 shows Hamming's method used to solve the equation $dy/dt = y$. This is the problem used in section 5.6.

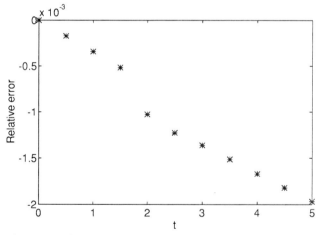

Fig. 5.7.1. Solution of $dy/dt = y$ where $y = 1$ when $t = 0$.
Relative error using Hamming's method with a step size of 0.5.

5.8 ERROR PROPAGATION IN DIFFERENTIAL EQUATIONS

In the preceding sections we have described various techniques for solving differential equations and the order, or a specific expression, for the truncation error at each step has been given. As we have discussed in section 5.3 for the Euler and trapezoidal methods, it is important to examine not only the magnitude of the error at each step but also how that error accumulates as the number of steps taken increases.

For the predictor–corrector method described above it can be shown that the predictor–corrector formulae introduce additional spurious solutions. As the iterative process proceeds, for some problems, the effect of these spurious solutions may be to overwhelm the true solution. In these circumstances the method is said to be unstable. Clearly we seek stable methods where the error does not develop in an unpredictable and unbounded way.

It is important to examine each numerical method to see if it is stable. In addition, if it is not stable for all differential equations we should provide tests to determine when it can be used with confidence. The theoretical study of stability for differential equations is a major undertaking and it is not intended to include a detailed analysis here. In section 5.9 we summarise the stability characteristics of specific methods and compare the performance of the major methods considered on a number of example differential equations.

5.9 THE STABILITY OF PARTICULAR NUMERICAL METHODS

A good discussion of the stability of many of the numerical methods for solving first-order differential equations is given by Ralston and Rabinowitz (1978) and Lambert (1973). Some of the more significant features, assuming all variables are real, are as follows.

(1) The Euler and trapezoidal methods: for a detailed discussion see sections 5.3 and 5.4.

(2) Runge–Kutta methods: Runge–Kutta methods do not introduce spurious solutions but instability may arise for some values of h. This may be removed by reducing h to a sufficiently small value. We have already described how the Runge–Kutta methods are less efficient than the predictor–corrector methods because of the greater number of function evaluations that may be required at each step by the Runge–Kutta methods. If h is reduced too far the number of function evaluations required may make the method uneconomic. The restriction on the size of the interval required to maintain stability may be estimated from the inequality $M < h\partial f/\partial y < 0$ where M is dependent on the particular Runge–Kutta method being used and may be estimated. Clearly this emphasises the need for careful step size adjustment during the solution process. This is efficiently implemented in the functions ode23 and ode45 so that this question does not present a problem when applying these MATLAB functions.

(3) Adams–Bashforth–Moulton method: in the *PECE* mode the range of absolute stability is given by $-1.25 < h\partial f/\partial y < 0$, implying that $\partial f/\partial y$ must be negative for absolute stability.

(4) Hamming's method: in the *PECE* mode the range of absolute stability is given by $-0.5 < h\partial f/\partial y < 0$, implying that $\partial f/\partial y$ must be negative for absolute stability.

Notice that the formulae given for estimating the step size can be difficult to use if f is a general function of y and t. However, in some cases the derivative of f is easily calculated: for example, when $f = Cy$ where C is a constant.

We now give some results of applying the methods discussed in previous sections to solve more general problems.

Example 1. Solve

$$dy/dt = 2yt \quad \text{where } y = 2 \text{ when } t = 0. \text{ Exact solution: } y = 2\exp(t^2).$$

We use the following script to obtain our results:

```
t0=0; tf=2; tinc=0.25; steps=floor((tf-t0)/tinc+1);
[t,x1]=abm('f503',t0,tf,2,tinc);
[t,x2]=fhamming('f503',t0,tf,2,tinc);
[t,x3]=rkgen('f503',t0,tf,2,tinc,1);
disp('Solution of dy/dt=2yt')
disp('t        abm       Hamming      Classical      Exact');
for i=1:steps
fprintf('%4.2f%12.7f%12.7f',t(i),x1(i),x2(i));
fprintf('%12.7f%12.7f\n',x3(i),2*exp(t(i)*t(i)));
end
```

The MATLAB function f503 is defined as

```
function v=f503(t,x)
v=2*t*x;
```

Running the above script produces the following output:

```
Classical method selected
Solution of dy/dt=2yt
t        abm       Hamming      Classical      Exact
0.00    2.0000000    2.0000000    2.0000000    2.0000000
0.25    2.1289876    2.1289876    2.1289876    2.1289889
0.50    2.5680329    2.5680329    2.5680329    2.5680508
0.75    3.5099767    3.5099767    3.5099767    3.5101093
1.00    5.4340314    5.4294215    5.4357436    5.4365637
1.25    9.5206761    9.5152921    9.5369365    9.5414664
1.50   18.8575896   18.8690552   18.9519740   18.9754717
1.75   42.1631012   42.2832017   42.6424234   42.7618855
2.00  106.2068597  106.9045567  108.5814979  109.1963001
```

Example 2. Solve

$$(1 + t^2)dy/dt + 2ty = \cos t \quad \text{where } y = 0 \text{ when } t = 0.$$

$$\text{Exact solution: } y = (\sin t)/(1 + t^2).$$

The script and function used to solve Example 1 are modified appropriately to solve the above equation and give the following output:

```
Classical method selected
Solution of (1+t^2)dy/dt=cost-2yt
t          abm          Hamming      Classical    Exact
0.00    0.0000000    0.0000000    0.0000000    0.0000000
0.25    0.2328491    0.2328491    0.2328491    0.2328508
0.50    0.3835216    0.3835216    0.3835216    0.3835404
0.75    0.4362151    0.4362151    0.4362151    0.4362488
1.00    0.4181300    0.4196303    0.4206992    0.4207355
1.25    0.3671577    0.3705252    0.3703035    0.3703355
1.50    0.3044513    0.3078591    0.3068955    0.3069215
1.75    0.2404465    0.2432427    0.2421911    0.2422119
2.00    0.1805739    0.1827267    0.1818429    0.1818595
```

Example 3. Solve

$$dy/dt = 3y/t \quad \text{where } y = 1 \text{ when } t = 1. \text{ Exact solution: } y = t^3.$$

The script and function used to solve Example 1 are again modified appropriately to solve the above equation and give the following output:

```
Classical method selected
Solution of dy/dt=3y/t
t          abm          Hamming      Classical    Exact
1.00    1.0000000    1.0000000    1.0000000    1.0000000
1.25    1.9518519    1.9518519    1.9518519    1.9531250
1.50    3.3719182    3.3719182    3.3719182    3.3750000
1.75    5.3538346    5.3538346    5.3538346    5.3593750
2.00    7.9916917    7.9919728    7.9912355    8.0000000
```

Examples 2 and 3 appear to show that there is little difference between the three methods considered and they are all fairly successful for the step size $h = 0.25$ in this range. Example 1 is a relatively difficult problem in which the classical Runge–Kutta method performs well.

5.10 SYSTEMS OF SIMULTANEOUS DIFFERENTIAL EQUATIONS

The numerical techniques we have described for solving a single first-order differential equation can be applied, after simple modification, to solve systems of first-order differential equations. Systems of differential equations arise naturally from mathematical models of the physical world. In this section we shall introduce a system of differential equations by considering a relatively simple example. This example is based on a much simplified model of the heart introduced by Zeeman and incorporates ideas from catastrophe theory. The model is described briefly here but more detail is given in the excellent text of Beltrami (1987). The resulting system of differential equations will be solved using the MATLAB toolbox function ode23 and the graphical facilities of MATLAB will help to clarify the interpretation of the results.

The starting point for this model of the heart is Van der Pol's equation which may be written in the form

$$dx/dt = u - \mu(x^3/3 - x)$$

$$du/dt = -x$$

This is a system of two simultaneous equations. The choice of this differential equation reflects our wish to imitate the beat of the heart. The fluctuation in the length of the heart fibre, as the heart contracts and dilates subject to an electrical stimulus, thus pumping blood through the system, may be represented by this pair of differential equations. The fluctuation has certain subtleties which our model should allow for. Starting from the relaxed state the contraction begins with the application of the stimulus slowly at first and then becomes faster so giving a sufficient final impetus to the blood. When the stimulus is removed the heart dilates slowly at first and then more rapidly until the relaxed state is again reached and the cycle can begin again.

To follow this behaviour the Van der Pol equation requires some modification so that the x variable represents the length of heart fibre and the variable u can be replaced by one which represents the stimulus applied to the heart. This is achieved by substituting $s = -u/\mu$, where s represents the stimulus and μ is a constant. Since ds/dt is equal to $(-du/dt)/\mu$ it follows that $du/dt = -\mu ds/dt$. Hence we obtain

$$dx/dt = \mu(-s - x^3/3 + x)$$

$$ds/dt = x/\mu$$

If these differential equations are solved for s and x for a range of time values we would find that s and x oscillate in a manner representing the fluctuations in the heart fibre length and stimulus. However, Zeeman proposed the introduction into this model of a tension factor p, where $p > 0$, in an attempt to account for the effects of increased blood pressure in terms of increased tension on the heart fibre. The model he suggested has the form

$$dx/dt = \mu(-s - x^3/3 + px)$$

$$ds/dt = x/\mu$$

Although the motivation for such a modification is plausible the effects of these changes are by no means obvious.

This problem provides an interesting opportunity to apply MATLAB to simulate the heart beat in an experimental environment that allows us to monitor its changes under the effects of differing tension values. The following script solves the differential equations and draws various graphs.

```
% Solving Zeeman's Catastrophe model of the heart
global p
p=input('enter tension value');
simtime=input('enter runtime');
acc=input('enter accuracy value');
initx=[0 -1]';
[t x]=ode23('f504',0,simtime,initx,acc);
%Plot results against time
plot(t,x(:,1),'--',t,x(:,2),'-');
xlabel('time'); ylabel('x and s');
```

Function f504 is defined as follows:

```
function fv=f504(t,x)
%note that x and s are represented by x(1) and x(2)
global p
fv=zeros(2,1);
fv(1) = 0.5*(-x(2)-x(1)^3/3+p*x(1));
fv(2)=2*x(1);
```

In the above function definition $\mu = 0.5$. Fig. 5.10.1 shows graphs of the fibre length x and s against time for a relatively small tension factor set at 1. The graphs show a steady periodic oscillation of fibre length for this tension value is achieved for small stimulus values. However, Fig. 5.10.2 plots x and s against time with the tension set at 20. This shows that the behaviour of the oscillation is clearly more laboured and much larger values of stimulus are required to produce the fluctuations in fibre length for the much higher tension value. Thus the graphs show the deterioration in the beat with increasing tension. The results parallel the expected physical effects and also give some degree of experimental support to the validity of this simple model.

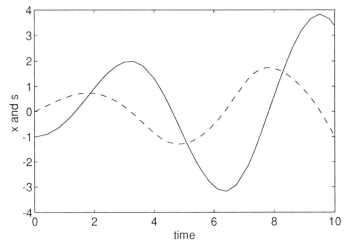

Fig. 5.10.1. Solution of Zeeman's model for $p = 1$ and accuracy 0.005.
The solid line represents s and the dashed line represents x.

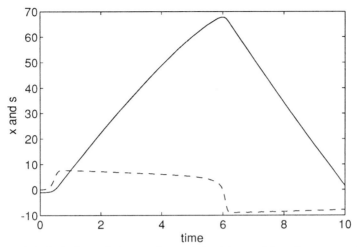

Fig. 5.10.2. Solution of Zeeman's model with $p = 20$ and accuracy 0.005.
The solid line represents s and the dashed line represents x.

A further interesting study can be made. The interrelation of the three parameters x, s and p can be represented by a three-dimensional surface called the cusp catastrophe surface. This surface can be shown to have the form

$$-s - x^3/3 + px = 0$$

See Beltrami (1987) for a more detailed explanation. Fig. 5.10.3 shows a series of sections of the cusp catastrophe curve for $p = 0:10:40$.

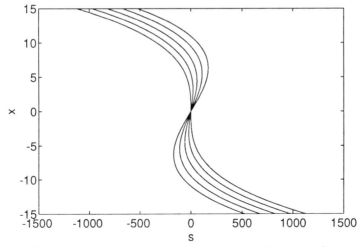

Fig. 5.10.3. Sections of the cusp catastrophe curve in
Zeeman's model for $p = 0:10:40$.

The curve has a pleat which becomes increasingly pronounced in the direction of increasing p. High tension or high p value consequently corresponds to movement on the sharply pleated part of this surface and consequently provides smaller changes in the heart fibre length relative to the stimulus.

5.11 THE LORENZ EQUATIONS

As an example of a system of three simultaneous equations we consider the Lorenz system. This system has a number of important applications including weather forecasting. The system has the form

$$dx/dt = s(y - x)$$

$$dy/dt = rx - y - xz$$

$$dz/dt = xy - bz$$

subject to appropriate initial conditions. As the parameters s, r and b are varied through various ranges of values the solutions of this system of differential equations vary in form. In particular for certain values of the parameters the system exhibits chaotic behaviour. To provide more accuracy in the computation process we use the MATLAB toolbox function ode45. The MATLAB script for solving this problem is given below.

```
%Solution of the Lorenz equation
global r
r=input('enter a value for the constant r');
simtime=input('enter runtime');
acc=input('enter accuracy value');
initx=[-7.69 -15.61 90.39]';
%Call ode45 to solve equations
[t x]=ode45('f505',0,simtime,initx,acc);
%Plot results against time
figure(1); plot(t,x);
xlabel('time'); ylabel('x');
figure(2); plot(x(:,1),x(:,3));
xlabel('x'); ylabel('z');
```

This script calls the function f505 which is defined thus:

```
function fv=f505(t,x)
%x, y and z are represented by x(1), x(2) and x(3)
global r
fv=zeros(3,1); fv(1)=10*(x(2)-x(1));
fv(2)=r*x(1)-x(2)-x(1)*x(3); fv(3)=x(1)*x(2)-8*x(3)/3;
```

The results of running this script are given in Figs 5.11.1 and 5.11.2. Fig. 5.11.1 is characteristic of the Lorenz equations and shows the complexity of the relationship between x and z. Fig. 5.11.2 shows how x, y and z change with time.

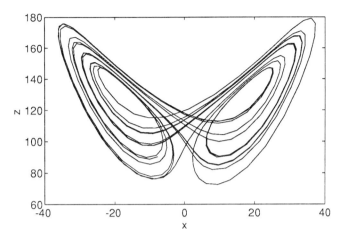

Fig. 5.11.1. Solution of Lorenz equations for $r = 126.52$,
using an accuracy of 0.000005 and terminating at $t = 8$.

When $r = 126.52$ and for other large values of r the behaviour of this system is chaotic. In fact, for $r > 24.7$ most orbits exhibit chaotic wandering. The trajectory passes around two points of attraction, called "strange attractors", switching from one to another in an apparently unpredictable fashion. This appearance of apparently random behaviour is remarkable considering the clearly deterministic nature of the problem. However, for other values of r the behaviour of the trajectories is simple and stable.

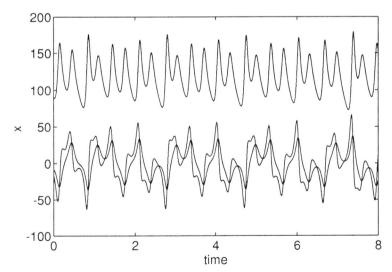

Fig. 5.11.2. Solution of Lorenz equations where each variable is plotted against time. Conditions are the same as those used to generate Fig. 5.11.1. Note the unpredictable nature of the solutions.

5.12 THE PREDATOR–PREY PROBLEM

A system of differential equations which models the interaction of competing or predator–prey populations is based on the Volterra equations and may be written in the form

$$dP/dt = K_1 P - CPQ$$

$$dQ/dt = -K_2 Q + DPQ$$

$$(5.12.1)$$

together with the initial conditions

$$Q = Q_0 \text{ and } P = P_0 \text{ at time } t = 0.$$

The variables P and Q give the size of the two interacting and competing populations at time t. Here K_1, K_2, C and D are positive constants. K_1 relates to the rate of growth of the prey population P and K_2 relates to the rate of decay of the predator population Q. It seems reasonable to assume that the number of encounters of predator and prey is proportional to P multiplied by Q and that a proportion C of these encounters will be fatal to members of the prey population. Thus the term CPQ gives a measure of the decrease in the prey population and the unrestricted growth in this population, which could occur assuming ample food, must be modified by the subtraction of this term. Similarly the decrease in the population of the predator must be modified by the addition of the term DPQ since the predator population gains food from its encounters with its prey and therefore more of the predators survive.

The solution of the differential equation depends on the specific values of the constants and will often result in nature in a stable cyclic variation of the populations. This is because as the predators continue to eat the prey, the prey population will fall and become insufficient to support the predator population which itself then falls. However, as the predator population falls more of the prey survive and consequently the prey population will then increase. This in turn leads to an increase in the predator population since it then has more food and the cycle begins again. This cycle maintains the predator–prey populations between certain upper and lower limits. The Volterra differential equations can be solved directly but this solution does not provide a simple relation between the size of the predator and prey populations and therefore numerical methods of solution should be applied. An interesting description of this problem is given by Simmons (1972).

We now use MATLAB to study the behaviour of a system of equations of the form (5.12.1) applied to the interaction of the lynx and its prey, the hare. The choice of the constants K_1, K_2, C and D is not a simple matter if we wish to obtain a stable situation where the populations of the predator and prey never die out completely but oscillate between upper and lower limits. The MATLAB script below uses $K_1 = 2$, $K_2 = 10$, $C = 0.001$ and $D = 0.002$ and considers the interaction of a population of lynxes and hares where it is assumed that this interaction is the crucial feature in determining the size of the two populations. With an initial population of 5000 hares and 100 lynxes the script uses these values to produce the graph in Fig. 5.12.1.

```
% x(1) is the hare population and x(2) the lynx population
simtime=input('enter runtime');
acc=input('enter accuracy value');
%Initialise values of populations
initx=[5000 100]';
[t x]=ode23('f506',0,simtime,initx,acc);
plot(t,x);
xlabel('time'); ylabel('population');
```

The function f506 is defined below:

```
function fv=f506(t,x)
fv=zeros(2,1);
fv(1)=2*x(1)-0.001*x(1)*x(2);
fv(2)=-10*x(2)+0.002*x(1)*x(2);
```

For these parameters there is a remarkably wide variation in the populations of the hares and lynxes. The lynx population, although periodically small, still recovers following a recovery of the hare population.

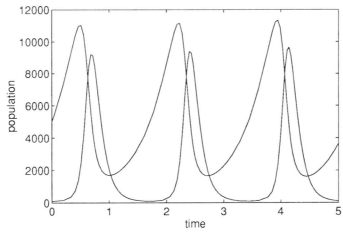

Fig. 5.12.1. Variation in the population of lynxes and hares against time, beginning with 5000 hares and 100 lynxes. Accuracy = 0.005.

5.13 DIFFERENTIAL EQUATIONS APPLIED TO NEURAL NETS

Different types of neural networks have been used to solve a wide range of problems. Neural nets often consist of several layers of "neurons" that are "trained" by fixing a set of weights. These weights are found by minimising the sum of squares of the difference between actual and required outputs. Once trained, the networks can be used to classify a range of inputs. However, here we consider a different approach that uses a neural network that may be based directly on considering a system of differential equations. This approach is described by Hopfield and Tank (1985, 1986) who demonstrated the application of neural networks to solving specific numerical problems. It is not our intention to provide the full details or proofs of this process here.

Hopfield and Tank, in their 1985 and 1986 papers, utilised a system of differential equations which take the form

$$\frac{\mathrm{d}u_i}{\mathrm{d}t} = \frac{-u_i}{\tau} + \sum_{j=0}^{n} T_{ij}V_j + I_i \quad \text{for } i = 0, 1, ..., n \tag{5.13.1}$$

where τ is a constant usually taken as 1. This system of differential equations represents the interaction of a system of n neurons and each differential equation is a simple model of a single biological neuron. (This is only one of a number of possible models of a neural network.) Clearly to establish a network of such neurons they must be able to interact with each other and this interaction must be represented in the differential equations. The T_{ij} provide the strengths of the interconnections between each neuron and the I_i provide the externally applied current to each neuron. These I_i may be viewed as inputs to the system. The V_j values provide the outputs from the system and are directly related to the u_j so that we may write $V_j = g(u_j)$. The function g, called a sigmoidal function, may be specified, for example, by

$$V_j = (1 + \tanh u_j)/2 \quad \text{for all} \quad j = 0, 1, ..., n$$

A plot of this function is given in Fig. 5.13.1.

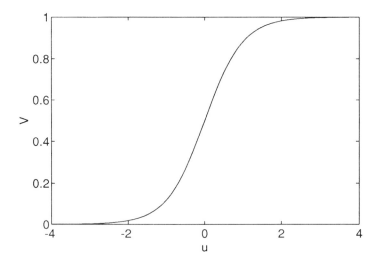

Fig. 5.13.1. Plot of sigmoid function $V = (1 + \tanh u)/2$.

Having provided such a model of a neural network the question still remains: how can we show that it can be used to solve specific problems? This is the key issue and a significant problem in itself. Before we can solve a given problem using a neural network we must first reformulate our problem so that it can be solved by this approach.

To illustrate this process Hopfield and Tank chose as an example the simple problem of binary conversion. That is, to find the binary equivalent of a given decimal number. Since there is no obvious and direct relationship between this problem and the system of differential equations (5.13.1) which model the neural network a more direct link has to be established.

Hopfield and Tank have shown that the stable state solution of (5.13.1), in terms of the V_j, is given by the minima of the energy function:

$$E = -\frac{1}{2} \sum_{i=0}^{n} \sum_{j=0}^{n} T_{ij} V_i V_j - \sum_{j=0}^{n} I_j V_j \qquad (5.13.2)$$

It is an easier matter to link the solution of the binary conversion problem to the minimisation of the function (5.13.2).

Hopfield and Tank consider the energy function

$$E = \frac{1}{2} \left| x - \sum_{j=0}^{n} V_j 2^j \right|^2 + \sum_{j=0}^{n} 2^{2j-1} V_j (1 - V_j) \qquad (5.13.3)$$

Now the minimum of (5.13.3) will be attained when $x = \sum V_j 2^j$ and $V_j = 0$ or 1. Clearly the first term ensures that the required binary representation is achieved while the second term provides that the V_j take either 0 or 1 values when the value of E is minimised. On comparing this energy function (5.13.3) with the general energy function (5.13.2) we find that if we make

$$T_{ij} = -2^{i+j} \text{ for } i \neq j \text{ and } T_{ij} = 0 \text{ when } i = j$$

$$I_j = -2^{2j-1} + 2^j x$$

then the two energy functions are equivalent, apart from a constant. Thus the minimum of one gives the minimum of the other. Thus solving the binary conversion problem expressed in this way is equivalent to solving the system of differential equations (5.13.1) with this special choice of values for T_{ij} and I_j.

In fact by using an appropriate choice of I_i and T_{ij} a range of problems can be represented by a neural network in the form of the system of differential equations (5.13.1). Hopfield and Tank have extended this process from the simple example considered above to attempting to solve the very challenging travelling salesman problem. The details of this are given in Hopfield and Tank (1985, 1986).

In MATLAB we may use ode23 or ode45 to solve this problem. The crucial part of this exercise is to define the function which gives the right-hand sides of the differential equation system for the neural network. This can be done very simply using MATLAB to write the function script below. This function gives the right-hand side for the differential equations which solve the binary conversion problem. In the definition of function f507, sc is the decimal value we wish to convert.

```
function neurf=f507(t,x)
global n sc
%Calculate synaptic current
I=2 .^[0:n-1]*sc-0.5*2 .^(2 .*[0:n-1]);
%Perform sigmoid transformation
V=(tanh(x/0.02)+1)/2;
%Compute interconnection values
p=2 .^[0:n-1].*V';
%Calculate change for each neuron
neurf=-x-2 .^[0:n-1]'*sum(p)+I'+2 .^(2 .*[0:n-1])'.*V;
```

This function f507 is called by the script below to solve the system of differential equations which define the neural network and hence simulate its operation.

```
% Hopfield and Tank neuron model for binary conversion problem
global n sc
n=input('enter number of neurons');
sc=input('enter number to be converted to binary form');
simtime=input('enter runtime');
acc=input('enter accuracy value');
initx=zeros(1,n)';
%Call ode45 to solve equation
[t x]=ode45('f507',0,simtime,initx,acc);
plot(t,(tanh(x/.02)+1)/2);
xlabel('time');ylabel('V');
```

The results of running this script to obtain the binary conversion for the decimal number 5 are shown in Fig. 5.13.2. This plot shows how the neural network model converges to the required results, i.e. $V(1) = 1$, $V(2) = 0$ and $V(3) = 1$ or the binary number 101.

This is an application of neural networks to a trivial problem. A real test for neural computing is the travelling salesman problem.

The MATLAB neural network toolbox provides a range of functions to solve neural network problems.

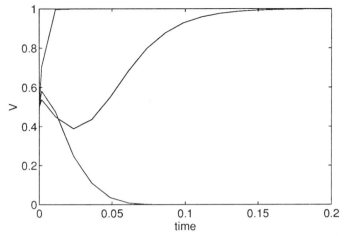

Fig. 5.13.2. Neural network finds the binary equivalent of 5
using three neurons and an accuracy of 0.005. The three curves
show the convergence to the binary digits 1, 0 and 1.

5.14 HIGHER-ORDER DIFFERENTIAL EQUATIONS

Higher-order differential equations can be solved by converting them to a system of
first-order differential equations. For example, to illustrate this consider the second-
order differential equation

$$2d^2x/dt^2 + 4(dx/dt)^2 - 2x = \cos x \qquad\qquad (5.14.1)$$

together with the initial conditions

$$x = 0 \text{ and } dx/dt = 10 \text{ when } t = 0$$

If we substitute $p = dx/dt$ then (5.14.1) becomes

$$2dp/dt + 4p^2 = \cos x + 2x$$

$$\qquad\qquad (5.14.2)$$

$$dx/dt = p$$

with initial conditions

$$p = 10 \text{ and } x = 0 \text{ when } t = 0$$

The second-order differential equations have been replaced by a system of first-order
differential equations. If we have an nth order differential equation of the form

$$a_n d^n y/dt^n + a_{n-1} d^{n-1} y/dt^{n-1} + \dots + a_0 y = f(t, y) \qquad (5.14.3)$$

by making the substitutions

$$P_0 = y \text{ and } dP_{i-1}/dt = P_i \quad \text{for } i = 1, 2, \dots, n-1 \qquad (5.14.4)$$

then (5.14.3) becomes

$$a_n dP_{n-1}/dt = f(t, y) - a_{n-1} P_{n-1} - a_{n-2} P_{n-2} - \dots - a_0 P_0 \qquad (5.14.5)$$

Now equations (5.14.4) and (5.14.5) together constitute a system of n first-order differential equations. Initial values will be given for (5.14.3) in terms of the various order derivatives P_i for $i = 1, 2, \dots, n-1$ at some initial value t_0 and these can easily be translated into initial conditions for the system of equations (5.14.4) and (5.14.5). In general the solutions of the original nth order differential equation and the system of first-order differential equations, (5.14.4) and (5.14.5), are the same. In particular the numerical solution will provide the values of y for a specified range of t. An excellent discussion of the equivalence of the solutions of the two problems is given in Simmons (1972). We can see from this description that any order differential equation of the form (5.14.3) with given initial values can be reduced to solving a system of first-order differential equations. This argument is easily extended to the more general nth order differential equation by making exactly the same substitutions as above in

$$d^n y/dt^n = f(t, y, y', \dots, y^{(n-1)})$$

where $y^{(n-1)}$ denotes the $(n-1)$th order derivative of y.

5.15 STIFF EQUATIONS

When the solution of a differential equation contains components which change at significantly different rates for given changes in the independent variable the equation is said to be "stiff". A differential equation or a system of differential equations may be affected by this phenomenon and when it is present a particularly careful choice of the step size must be made if stability is to be achieved.

We will now consider how the stiffness phenomenon arises in an apparently simple system of differential equations. Consider the system given below.

$$dy_1/dt = -by_1 - cy_2$$
$$\qquad (5.15.1)$$
$$dy_2/dt = y_1$$

This system may be written in matrix form as

$$dy/dt = \mathbf{A}y \qquad (5.15.2)$$

The solution of this system is

$$y_1 = A \exp(r_1 t) + B \exp(r_2 t)$$

$$y_2 = C \exp(r_1 t) + D \exp(r_2 t)$$

(5.15.3)

where A, B, C and D are constants set by the initial conditions. It can be easily verified that r_1 and r_2 are the eigenvalues of the matrix **A**.

By taking various values of b and c in (5.15.1) we can generate many problems of the form (5.15.2) having solutions (5.15.3) where the eigenvalues r_1 and r_2 will, of course, change from problem to problem. If a numerical procedure is applied to solve these systems of differential equations the success of the method will depend crucially on the eigenvalues of the matrix **A** and in particular the ratio of the smallest and largest eigenvalues.

The purpose of the script below is to investigate how the difficulty of solving (5.15.1) depends on the ratio of the largest and smallest eigenvalues by comparing the time taken to solve specific problems.

```
global a
b=[3 20 100 500]; c=[2 .1 1 1];
for i=1:4
  a=[-b(i) -c(i);1 0];
  lambda=eig(a);
  eigenratio(i)=max(abs(lambda))/min(abs(lambda));
  time0=clock;
  inity=[0 1]';
  [t,y]=ode23('f508',0,2,inity,.005);
  et(i)=etime(clock,time0);
end;
eigenratio
et
```

This script calls the function f508 which gives the right-hand sides of the differential equations defined by (5.15.1):

```
function v=f508(t,y)
global a
v=a*y;
```

Running this script gives

```
eigenratio =
  1.0e+05 *
    0.0000    0.0400    0.1000    2.5000

et =
    1.4167    1.4667    8.1667    49.7167
```

The first eigenratio is 2, not zero. It is seen that as the eigenvalue ratio increases so does the time taken to solve the problem. Problems will arise if there is a wide variation in the magnitude of the eigenvalues.

As an example of a matrix with widely spaced eigenvalues we can take the Rosser matrix; this is available in MATLAB as `gallery(8)`. The sequence of statements

```
a=gallery(8);
lambda=eig(a);
eigratio=max(abs(lambda))/min(abs(lambda))
```

produces a matrix with eigenvalue ratios of order 10^{16}. Thus a system of ordinary first-order differential equations involving this matrix would be pathologically difficult. The significance of the eigenvalue ratio in relation to the required step size can be generalised to systems of many equations. Consider the system of n equations

$$dy/dt = Ay + p(t) \qquad (5.15.4)$$

where y is an n component column vector, $p(t)$ is an n component column vector of functions of t and A is an $n \times n$ matrix of constants. It can be shown that the solution of this system takes the form

$$y(t) = \sum_{i=1}^{n} v_i \, d_i \exp(r_i t) + s(t) \qquad (5.15.5)$$

Here r_1, r_2, ... are the eigenvalues and d_1, d_2, ... the eigenvectors of A. The vector function $s(t)$ is the particular integral of the system, sometimes called the steady state solution since for negative eigenvalues the exponential terms should die away with increasing t. If it is assumed that the $r_k < 0$ for $k = 1, 2, 3, ...$ and we require the steady state solution of system (5.15.4) then any numerical method applied to solve this problem may face significant difficulties as we have seen. We must continue the integration until the exponential components have been reduced to negligible levels and yet we must take sufficiently small steps to ensure stability, thus requiring many steps over a large interval. This is the most significant effect of stiffness.

The definition of stiffness can be extended to any system of the form (5.15.4). The stiffness ratio is defined as the ratio of the largest and smallest eigenvalues of A and gives a measure of the stiffness of the system.

The methods used to solve stiff problems must be based on stable techniques. The MATLAB functions ode23 and ode45 use continuous step size adjustment and therefore are able to deal with such problems although the solution process may be slow. If we use a predictor–corrector method, not only must this method be stable but the corrector must be iterated to convergence. An interesting discussion of this topic is given by Ralston and Rabinowitz (1978). Specialised methods have been developed for solving stiff problems and Gear (1971) has provided a number of techniques which have been reported to be successful.

5.16 SPECIAL TECHNIQUES

A further set of predictor–corrector equations may be generated by making use of an interpolation formula due to Hermite. An unusual feature of these equations is that they contain second-order derivatives. It is usually the case that the calculation of second-order derivatives is not particularly difficult and consequently this feature does not add a significant amount of work to the solution of the problem. However, it should be noted that in using a computer program for this technique the user has to supply not only the function on the right-hand side of the differential equation but its derivative as well. To the general user this may be unacceptable.

The equations for this method take the form

$$y_{n+1}^{(1)} = y_n + h(y'_n - 3y'_{n-1})/2 + h^2(17y''_n + 7y''_{n-1})/12$$

$$y*_{n+1}^{(1)} = y_{n+1}^{(1)} + 31(y_n - y_n^{(1)})/30 \qquad\qquad (5.16.1)$$

$$y'_{n+1}^{(1)} = f(t_{n+1}, y*_{n+1}^{(1)})$$

For $k = 1, 2, 3, \dots$

$$y_{n+1}^{(k+1)} = y_n + h(y'_{n+1}^{(k)} + y'_n)/2 + h^2(-y''_{n+1}^{(k)} + y''_n)/12$$

This method is stable and has a smaller truncation error at each step than Hamming's method. Thus it may be worthwhile accepting the additional effort required by the user. The method is known as the Hermite method. We note that since we have

$$dy/dt = f(t, y)$$

then

$$d^2y/dt^2 = df/dt$$

and thus y''_n etc. are easily calculated as the first derivative of f. The MATLAB function fhermite implements this method and the script is given below. Note that in this function, the function f must provide both the first and second derivatives of y.

```
function [tvals, yvals]=fhermite(f,start,finish,startval,step)
% Solves dx/dt=f(t,x). start, finish are initial, final values of t
% startval is initial value of x, step is the increment in t
% 3 steps of Runge-Kutta are required so that hermite can start.
% Set up matrices for Runge-Kutta methods
b=[ ];c=[ ];d=[ ];
order=4;
b=[1/6 1/3 1/3 1/6]; d=[0 0.5 0.5 1];
c=[0 0 0 0;0.5 0 0 0;0 0.5 0 0;0 0 1 0];
steps=(finish-start)/step+1;
y=startval; t=start;
ys(1)=startval; [fval(1),df(1)]=feval(f,t,y);
yvals=startval; tvals=start;                          [Script continues...
```

```
for j=2:2
  k(1)=step*fval(1);
  for i=2:order
    k(i)=step*feval(f,t+step*d(i),y+c(i,1:i-1)*k(1:i-1)');
  end;
  y1=y+b*k'; ys(j)=y1; t1=t+step;
  [fval(j),df(j)]=feval(f,t1,y1);
  %collect values together for output
  tvals=[tvals, t1]; yvals=[yvals, y1];
  t=t1; y=y1;
end;
%hermite now applied
h2=step*step/12; er=1;
for i=3:steps
  y1=ys(2)+step*(3*fval(1)-fval(2))/2+h2*(17*df(2)+7*df(1));
  t1=t+step; y1m=y1; y10=y1;
  if i>3, y1m=y1+31*(ys(2)-y10)/30; end;
  [fval(3), df(3)]=feval(f,t1,y1m);
  yc=0; er=1;
  while abs(er)>.0000001
    yp=ys(2)+step*(fval(2)+fval(3))/2+h2*(df(2)-df(3));
    [fval(3), df(3)]=feval(f,t1,yp);
    er=yp-yc; yc=yp;
  end;
  fval(1)=fval(2); df(1)=df(2); fval(2)=fval(3); df(2)=df(3);
  ys(2)=yp;
  tvals=[tvals, t1]; yvals=[yvals, yp];
  t=t1;
end;
```

Fig. 5.16.1 gives the error when solving the specific equation $dy/dt = y$ using the same step size and starting point as for Hamming's method, see Fig. 5.7.1. For this particular

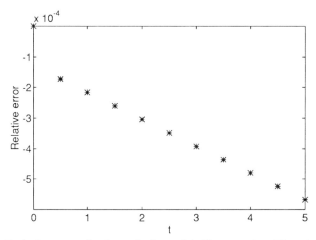

Fig. 5.16.1. Relative error in the solution of $dy/dt = y$ using Hermite's method.
Initial condition $y = 1$ when $t = 0$ and a step of 0.5.

problem Hermite's method performs better than Hamming's method.

Finally we compare the Hermite, Hamming and Adams–Bashforth–Moulton methods for the difficult problem

$$dy/dt = -10y \text{ given } y = 1 \text{ when } t = 0.$$

For the function fhermite we must supply both the first and second derivatives of y with respect to t. For the first derivative, we have directly $dy/dt = -10y$ but the second derivative d^2y/dt^2 is given by $-10dy/dt = -10(-10y) = 100y$. Consequently the function f509 takes the form

```
function [v, dv]=f509(t,x)
v=-10*x; dv=100*x;
```

The functions abm and fhamming require only the first derivative of y with respect to t and we define function f510 thus:

```
function v=f510(t,x)
v=-10*x;
```

The following results demonstrate the superiority of the Hermite method.

```
Solution of dx/dt=-10*x
t        abm          Hamming       Hermite       Exact
0.0    1.0000000    1.0000000     1.0000000     1.0000000
0.1    0.3750000    0.3750000     0.3750000     0.3678794
0.2    0.1406250    0.1406250     0.1381579     0.1353353
0.3    0.0527344    0.0527344     0.0509003     0.0497871
0.4   -0.0032654    0.0109440     0.0187528     0.0183156
0.5   -0.0171851    0.0070876     0.0069089     0.0067379
0.6   -0.0010598    0.0131483     0.0025454     0.0024788
0.7    0.0023606    0.0002607     0.0009378     0.0009119
0.8   -0.0063684    0.0006066     0.0003455     0.0003355
0.9   -0.0042478    0.0096271     0.0001273     0.0001234
1.0    0.0030171   -0.0065859     0.0000469     0.0000454
```

One feature which may be used to improve many of the methods discussed above is step size adjustment. This means that we adjust the step size h according to the progress of the iteration. One criterion for adjusting h is to monitor the size of the truncation error. If the truncation error is smaller than the accuracy requirement we can increase h but if the truncation error is too large we can reduce h. Step size adjustment can lead to considerable additional work; for example, if a predictor–corrector method is used new initial values must be calculated. The following method is an interesting alternative to this kind of procedure.

5.17 EXTRAPOLATION TECHNIQUES

This extrapolation method is based on a similar procedure to that used in Romberg integration, introduced in Chapter 4. The procedure begins by obtaining successive initial approximations for y_{n+1} using a modified mid-point method. The interval sizes used for obtaining these approximations are calculated from

$$h_i = h_{i-1}/2 \text{ for } i = 1, 2, 3, ... \tag{5.17.1}$$

with the initial value h_0 given.

Once these initial approximations have been obtained we can use (5.17.2), the extrapolation formula, to obtain improved approximations.

$$T_{m,k} = (4^m T_{m-1,k+1} - T_{m-1,k})/(4^m - 1) \tag{5.17.2}$$

$$\text{for } m = 1, 2, ..., s \text{ and } k = 0, 1, 2, ..., s - m$$

The calculations are set out in an array in much the same way as the calculations for Romberg's method for integration described in Chapter 4. When $m = 0$ the values of $T_{0,k}$ for $k = 0, 1, 2, ..., s$ are taken as the successive approximations to the values of y_{n+1} using the h_i values obtained from (5.17.1).

The formula for calculating the approximations used for the initial values $T_{0,k}$ in the above array are computed by using the equations (5.17.3).

$$y_1 = y_0 + hy'_0$$

$$y_{n+1} = y_{n-1} + 2hy'_n \quad \text{for } n = 1, 2, ..., N_k \tag{5.17.3}$$

Here $k = 1, 2, ...$ and N_k is the number of steps taken in the range of interest so that $N_k = 2^k$ as the size of the interval is halved each time. The distance $2h$ between y_{n+1} and y_{n-1} values may lead to significant variations in the magnitude of the error. Because of this, instead of using the final value of y_{n+1} given by (5.17.3), Gragg (1965) has suggested that at the final step these values are smoothed using the intermediate value y_n. This leads to the values for $T_{0,k}$

$$T_{0,k} = (y^k_{N-1} + 2y^k_N + y^k_{N+1})/4$$

where the superscript k denotes the value at the kth division of the interval.

Alternatives to the method of Gragg are available for finding the initial values in the table and various combinations of predictor–correctors may be used. It should be noted, however, that if the corrector is iterated until convergence is achieved this will improve the accuracy of the initial values but at considerable computational expense for smaller step sizes, i.e. for larger N values. The MATLAB function rombergx implements the extrapolation method and is given below.

```
function v=rombergx(f,start,finish,intdiv,inity)
% Solves dy/dt=f(t,y). start, finish are initial, final values of t
% inity is initial value of y
% intdiv is the number of interval divisions required.
% Calculate initial values
for index=1:intdiv
  y0=inity; t0=start;
  intervals=2^index;
  step=(finish-start)/intervals;
  y1=y0+step*feval(f,t0,y0);
  t=t0+step;
  for i=1:intervals
    y2=y0+2*step*feval(f,t,y1);
    t=t+step;
    ye2=y2; ye1=y1; ye0=y0; y0=y1; y1=y2;
  end;
  tableval(index)=(ye0+2*ye1+ye2)/4;
end;
for i=1:intdiv-1
  for j=1:intdiv-i
    table(j)=(tableval(j+1)*4^i-tableval(j))/(4^i-1);
    tableval(j)=table(j);
  end;
  tablep=table(1:intdiv-i);
  disp(tablep)
end;
```

We can now call this function to solve $dx/dt = -10x$ at $x = 0.5$. Note that the right-hand side of this equation is defined by function f510 which is given in section 5.16.

The following MATLAB command solves this differential equation:

```
»rombergx('f510',0,0.5,7,1)
  -2.5677    0.2277    0.1624    0.0245    0.0080    0.0068
   0.4141    0.1580    0.0153    0.0069    0.0067
   0.1539    0.0131    0.0068    0.0067
   0.0125    0.0068    0.0067
   0.0068    0.0067
   0.0067
```

The final value, 0.0067, is better than any of the results achieved for this problem by other methods presented in this chapter. It must be noted that only the final value is found; other values in a given interval can be obtained if intermediate ranges are considered.

This completes our discussion of those types of differential equations known as initial value problems. In Chapter 6 we consider a different type of differential equation known as a boundary value problem.

5.18 SUMMARY

This chapter has introduced a range of MATLAB functions for solving differential equations and systems of differential equations which supplement ode23 and ode45 provided in the MATLAB toolbox. We have demonstrated how these functions may be used to solve a wide variety of problems.

PROBLEMS

5.1. A radioactive material decays at a rate that is proportional to the amount that remains. The differential equation which models this process is

$$dy/dt = -ky \text{ where } y = y_0 \text{ when } t = t_0$$

Here y_0 represents the mass at the time t_0. Solve this equation for $t = 0$ to 10 given that $y_0 = 50$ and $k = 0.05$, using

    (i) the function `feuler`, with $h = 1, 0.1, 0.01$.
    (ii) the function `eulertp`, with $h = 1, 0.1$.
    (iii) the function `rkgen`, set for the classical method, with $h = 1$.

Compare your results with the exact solution, $y = 50 \exp(-0.05t)$.

5.2. Solve $y' = 2xy$ with initial conditions $y_0 = 2$ when $x_0 = 0$ in the range $x = 0$ to 2. Use the classical, Merson and Butcher variants of the Runge–Kutta method, all implemented in function `rkgen`, with step $h = 0.2$. Note that the exact solution is $y = 2\exp(x^2)$.

5.3. Repeat problem 5.1 using the following predictor–corrector methods with $h = 2$ for $t = 0$ to 50:

    (i) Adams–Bashforth–Moulton's method, function `abm`.
    (ii) Hamming's method, function `fhamming`.

5.4. Express the following second-order differential equation as a pair of first-order equations:

$$xy'' - y' - 8x^3y^3 = 0$$

with the initial conditions $y = 1/2$ and $y' = -1/2$ at $x = 1$. Solve the pair of first-order equations using both ode23 and ode45 in the range 1 to 4. The exact solution is $y = 1/(1 + x^2)$.

5.5. Use function `fhermite` to solve

(i) problem 5.1 with $h = 1$.
(ii) problem 5.2 with $h = 0.2$.
(iii) problem 5.2 with $h = 0.02$.

5.6. Use the MATLAB function `rombergx` to solve the following problems. In each case use eight divisions.

(i) $y' = 3y/x$ with initial conditions $x = 1$, $y = 1$. Determine y when $x = 20$.
(ii) $y' = 2xy$ with initial conditions $x = 0$, $y = 2$. Determine y when $x = 2$.

5.7. Consider the predator–prey problem described in section 5.12. This problem may be extended to consider the effect of culling on the interacting populations by subtracting a term from both equations in (5.12.1) as follows:

$$dP/dt = K_1 P - CPQ - S_1 P$$
$$dQ/dt = -K_2 Q + DPQ - S_2 Q$$

Here S_1 and S_2 are constants which provide the culling level for the populations. Use `ode45` to solve this problem with $K_1 = 2$, $K_2 = 10$, $C = 0.001$ and $D = 0.002$ and initial values of the population $P = 5000$ and $Q = 100$. Assuming that S_1 and S_2 are equal, experiment with values in the range 1 to 2. There is a wealth of experimental opportunity in this problem and the reader is encouraged to investigate different values of S_1 and S_2.

5.8. Solve the Lorenz equations given in section 5.11 using function `f505` with `ode23` for $r = 1$.

5.9. Use the Adams–Bashforth–Moulton method to solve $dy/dt = -5y$, with $y = 50$ when $t = 0$, in the range $t = 0$ to 6. Try step sizes, h, of 0.1, 0.2, 0.25 and 0.4. Plot the error against t for each case. What can you deduce from these results with regard to the stability of the method? The exact answer is $y = 50e^{-5t}$.

6

Boundary value problems

6.1 INTRODUCTION

In Chapter 5 we have examined methods for solving initial value ordinary differential equations. The solution of these equations depended on the nature of the equation and the specified initial conditions. In this chapter the solution of boundary value problems and problems that have both boundary and initial values are examined. The solution of a boundary value problem in one independent variable must satisfy specified conditions at two points and the solution of a boundary value problem in two independent variables must satisfy specified conditions at all points along a curve or set of lines enclosing a specified region. Although not considered in this chapter, a further important boundary value problem is one with three independent variables, for example Laplace's equation in three dimensions. In this case the solution must satisfy specified conditions, all points over a surface enclosing a specified volume. It should be noted, of course, that in a mixed boundary and initial value problem one independent variable, usually time, will be associated with one or more initial values and the remaining independent variables will be associated with boundary values.

In this chapter we will restrict the discussion to second-order differential equations in one or two independent variables. The general form of these equations for one and two independent variables is given by (6.1.1) and (6.1.2) respectively.

$$A(x)\frac{d^2z}{dx^2} + f\left(x, z, \frac{dz}{dx}\right) = 0 \qquad (6.1.1)$$

$$A(x,y)\frac{\partial^2 z}{\partial x^2} + B(x,y)\frac{\partial^2 z}{\partial x\,\partial y} + C(x,y)\frac{\partial^2 z}{\partial y^2} + f\left(x, y, z, \frac{\partial z}{\partial x}, \frac{\partial z}{\partial y}\right) = 0 \qquad (6.1.2)$$

These equations are linear in the second-order terms but the terms

$$f\left(x, z, \frac{dz}{dx}\right) \text{ and } f\left(x, y, z, \frac{\partial z}{\partial x}, \frac{\partial z}{\partial y}\right)$$

may be linear or non-linear. Fig. 6.1.1 shows how these equations may be classified. In particular, (6.1.2) is classified as an elliptic, parabolic or hyperbolic partial differential equation as follows:

If $B^2 - 4AC < 0$ the equation is elliptic.
If $B^2 - 4AC = 0$ the equation is parabolic.
If $B^2 - 4AC > 0$ the equation is hyperbolic.

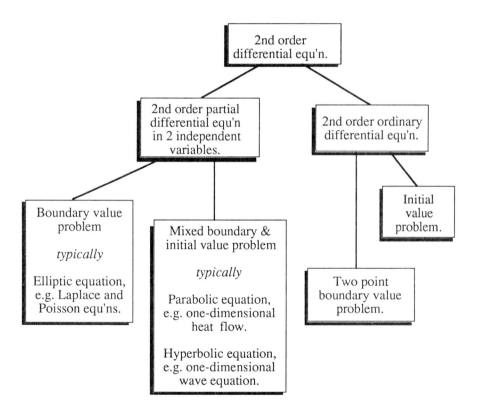

Fig. 6.1.1. Second-order differential equations with one
or two independent variables and their solutions.

Since the coefficients A, B and C are, in general, functions of the independent variables, the classification of (6.1.2) may vary in different regions of the domain in which the problem is defined. We will commence with a study of (6.1.1).

6.2 THE SHOOTING METHOD

An initial value problem and a two-point boundary value problem derived from the same differential equation may have the same solution. For example, consider the differential equation

$$\frac{d^2y}{dx^2} + y = \cos 2x \tag{6.2.1}$$

Given the initial conditions that when $x = 0$, $y = 0$ and $dy/dx = 1$, the solution of (6.2.1) is

$$y = (\cos x - \cos 2x)/3 + \sin x$$

However, this solution also satisfies (6.2.1) with the two boundary conditions $x = 0$, $y = 0$ and $x = 4/3$, $y = \pi/2$.

This observation provides a useful method of solving two-point boundary value problems – the "shooting method". As an example, consider the equation

$$x^2 \frac{d^2y}{dx^2} - 6y = 0 \tag{6.2.2}$$

with boundary conditions $y = 1$ when $x = 1$ and $x = 2$. We will treat this problem as an initial value problem where $y = 1$ when $x = 1$ and assume trial values for dy/dx when

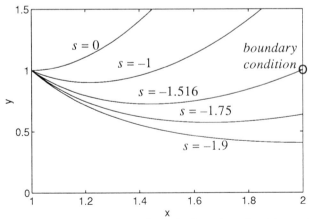

Fig. 6.2.1. Solutions for $x^2(d^2y/dx^2) - 6y = 0$ with $y = 1$
and $dy/dx = s$ when $x = 1$, for trial values of s.

$x = 1$, denoted by s. Fig. 6.2.1 shows the solution for various trial values of s. It is seen that when $s = -1.516$ the solution satisfies the required boundary condition that $y = 1$ when $x = 2$. The solution for (6.2.2) can be found by changing it into a pair of first-order differential equations and using any appropriate numerical method described in Chapter 5. Equation (6.2.2) is equivalent to

$$dy/dx = z$$

$$(6.2.3)$$

$$dz/dx = 6y/x^2$$

We must determine the slope dy/dx that gives the correct boundary condition. This could be achieved by trial and error but this is tedious and in practice we can use interpolation. The script below solves (6.2.3) for four trial slopes using the MATLAB function ode45. Vectors s contain trial values of the slope dy/dx at $x = 1$. Vector b contains the corresponding value of y when $x = 2$, computed by ode45. A slightly modified version of the function aitken (described in Chapter 7) which suppresses intermediate output is called to determine, by interpolation, the slope that gives the required boundary condition. Finally this slope, s0, is used in ode45 to determine the correct solution to (6.2.3).

```
s=-1.25:-0.25:-2; s0=[ ];
ncase=length(s);b=zeros(1,ncase);
for i=1:ncase;
   [x,y]=ode45('f601',1,2,[1 s(i)],.0005);
   [m,n]=size(y);
   b(1,i)=y(m,1);
end
s0=aitken1(b,s,1)
[x,y]=ode45('f601',1,2,[1 s0],.0005);
[x y(:,1)]
```

To define the right-hand sides of the differential equations (6.2.3) we use the function f601 given below.

```
function fv=f601(x,y)
%y and z are y(1) and y(2) respectively
fv=zeros(2,1);fv(1)=y(2);fv(2)=6*y(1)/x^2;
```

Running the above script gives

```
s0 =
   -1.5161

ans =
       1.0000    1.0000
       1.0078    0.9883
       1.0703    0.9071
       1.1328    0.8445
```

[Output continues...

1.1953	0.7974
1.2578	0.7635
1.3203	0.7409
1.3828	0.7282
1.4453	0.7246
1.5078	0.7290
1.5703	0.7410
1.6328	0.7601
1.6953	0.7858
1.7578	0.8179
1.8203	0.8563
1.8828	0.9007
1.9453	0.9511
2.0000	1.0000

The interpolated value of the slope is −1.5162. The first column of *ans* above gives the values of x and the second gives the corresponding values of y.

Whilst the shooting method is not particularly efficient, it does have the advantage of being able to solve non-linear boundary value problems. We will now examine an alternative method for solving boundary problems: the finite difference method.

6.3 THE FINITE DIFFERENCE METHOD

In Chapter 4 it is shown how derivatives can be approximated by the use of finite differences. We can use the same approach for the solution of certain types of differential equations. The method effectively replaces the differential equation by a set of approximate difference equations. The central difference approximations for the first and second derivatives of z with respect to x are given in (6.3.1) and (6.3.2) below. In these and subsequent equations the operator D_x represents d/dx, $D_x^2 = d^2/dx^2$, etc. The subscript x is omitted where there is no danger of confusion. Thus at z_i:

$$Dz_i \approx (-z_{i-1} + z_{i+1})/(2h) \tag{6.3.1}$$

$$D^2z_i \approx (z_{i-1} - 2z_i + z_{i+1})/h^2 \tag{6.3.2}$$

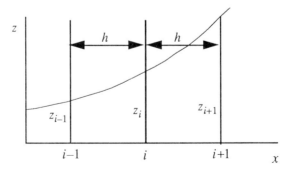

Fig. 6.3.1. Equally spaced nodal points.

In (6.3.1) and (6.3.2) h is the distance between the nodal points, see Fig. 6.3.1, and these approximating formulae have errors of order h^2. Higher-order approximations can be generated which have errors of order h^4, but we will not require them. To achieve the same degree of accuracy we will make h smaller.

We can also determine the approximations for unevenly spaced nodal points. For example, it can be shown that (6.3.1) and (6.3.2) become

$$Dz_i \approx \frac{1}{h\beta(\beta + 1)} \left\{ -\beta^2 z_{i-1} - \left(1 - \beta^2\right) z_i + z_{i+1} \right\} \tag{6.3.3}$$

$$D^2 z_i \approx \frac{2}{h^2 \beta(\beta + 1)} \left\{ \beta z_{i-1} - (1 + \beta) z_i + z_{i+1} \right\} \tag{6.3.4}$$

where $h = x_i - x_{i-1}$ and $\beta h = x_{i+1} - x_i$. Note that when $\beta = 1$, (6.3.3) and (6.3.4) simplify to (6.3.1) and (6.3.2) respectively. Approximation (6.3.3) has an error of order h^2, irrespective of the value of β, and (6.3.4) has an error of order h for $\beta \neq 1$ and h^2 for $\beta = 1$.

Equations (6.3.1) to (6.3.4) are central difference approximations, i.e. the approximation for a derivative uses values of the function either side of the point at which the derivative is to be determined. These are generally the most accurate approximations, but in some situations it is necessary to use forward or backward difference approximations. For example, the forward difference approximation for Dz_i is

$$Dz_i \approx (-z_i + z_{i+1})/h \quad \text{with an error of order } h \tag{6.3.5}$$

The backward difference approximation for Dz_i is

$$Dz_i \approx (-z_{i-1} + z_i)/h \quad \text{with an error of order } h \tag{6.3.6}$$

To determine solutions for partial differential equations we will require the finite difference approximation for various partial derivatives in two or more variables. These approximations can be derived by combining some of the above equations. For example, we can determine the finite difference approximation for $\partial^2 z/\partial x^2 + \partial^2 z/\partial y^2$ (i.e. $\nabla^2 z$) from the approximation (6.3.2) or (6.3.4). To avoid double subscripts we will use the notation applied to the mesh shown in Fig. 6.3.2. Thus, from (6.3.2),

$$\nabla^2 z_i \approx (z_l - 2z_i + z_r)/h^2 + (z_a - 2z_i + z_b)/k^2$$

$$\approx \{r^2 z_l + r^2 z_r + z_a + z_b - 2(1 + r^2) z_i\}/(r^2 h^2) \tag{6.3.7}$$

where $r = k/h$. If $r = 1$ then (6.3.7) becomes

$$\nabla^2 z_i \approx (z_l + z_r + z_a + z_b - 4z_i)/h^2 \tag{6.3.8}$$

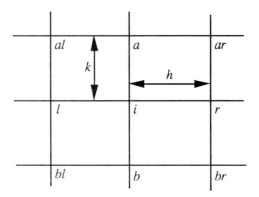

Fig. 6.3.2. Grid mesh in rectangular coordinates.

These central difference approximations for $\nabla^2 z_i$ have an error of $O(h^2)$.

The finite difference approximation for the second-order mixed derivative of z with respect to x and y, $\partial^2 z/\partial x \partial y$ or D_{xy}, is determined by applying (6.3.1) in the x direction to each term of (6.3.1) in the y direction thus:

$$D_{xy}z_i \approx [\{(z_r - z_l)_a\}/2h - \{(z_r - z_l)_b\}/2h]/2k$$

$$\approx (z_{ar} - z_{al} - z_{br} + z_{bl})/4hk \qquad (6.3.9)$$

We can develop the finite difference approximations in other coordinate systems such as skew and polar coordinates and we can have uneven spacing of the node points in any direction, see Salvadori and Baron (1961).

6.4 TWO-POINT BOUNDARY VALUE PROBLEMS

Before considering the application of finite difference methods to solve a differential equation we will first consider the nature of the solution. We will begin by considering the following second-order inhomogeneous differential equation in one independent variable:

$$(1 + x^2)\frac{d^2z}{dx^2} + x\frac{dz}{dx} - z = x^2 \qquad (6.4.1)$$

subject to the non-homogeneous boundary conditions $x = 0$, $z = 1$ and $x = 2$, $z = 2$. The solution of this equation is

$$z = -\frac{\sqrt{5}}{6}x + \frac{1}{3}(1 + x^2)^{1/2} + \frac{1}{3}(2 + x^2) \qquad (6.4.2)$$

This is the only solution that satisfies both the equation and its boundary conditions. In contrast to this, consider the solution of the second-order homogeneous equation

$$x\,\frac{d^2z}{dx^2} + \frac{dz}{dx} + \lambda x^{-1}z = 0 \tag{6.4.3}$$

subject to the homogeneous boundary conditions that $z = 0$ at $x = 1$ and $dz/dx = 0$ at $z = e$ (i.e. 2.7183...). If λ is a given constant this homogeneous equation has the trivial solution $z = 0$. However, if λ is an unknown then we can determine values of λ to give non-trivial solutions for z. Equation (6.4.3) is then a characteristic value or eigenvalue problem. Solving (6.4.3) gives an infinite number of solutions for λ and z as follows:

$$z_n = \sin\left\{(2n + 1)\frac{\pi}{2}\log_e|x|\right\},\quad \lambda_n = \{(2n + 1)\pi/2\}^2 \quad n = 0, 1, 2, ... \tag{6.4.4}$$

The values of λ that satisfy (6.4.3) are called characteristic values or eigenvalues, and the corresponding values of z are called characteristic functions or eigenfunctions. This particular type of boundary value problem is called a characteristic value or eigenvalue problem. It has arisen because both the differential equation and the specified boundary conditions are homogeneous.

The application of finite differences to the solution of boundary value problems is now illustrated by two examples.

Example 1. Determine an approximate solution for (6.4.1). We will begin by multiplying (6.4.1) by $2h^2$ and writing d^2z/dx^2 by D^2z, etc. thus:

$$2(1 + x^2)(h^2D^2z) + xh(2hDz) - 2h^2z = 2h^2x^2 \tag{6.4.5}$$

Using (6.3.1) and (6.3.2) we can replace (6.4.5) by

$$2(1 + x_i^2)(z_{i-1} - 2z_i + z_{i+1}) + x_ih(-z_{i-1} + z_{i+1}) - 2h^2z_i = 2h^2x_i^2 \tag{6.4.6}$$

Fig. 6.4.1 shows x divided into four segments ($h = 1/2$) with nodes numbered $1 - 5$. Applying (6.4.6) to nodes 2, 3 and 4 gives

At node 2: $2(1 + 0.5^2)(z_1 - 2z_2 + z_3) + 0.25(-z_1 + z_3) - 0.5z_2 = 0.5(0.5^2)$

At node 3: $2(1 + 1^2)(z_2 - 2z_3 + z_4) + 0.5(-z_2 + z_4) - 0.5z_3 = 0.5(1^2)$

At node 4: $2(1 + 1.5^2)(z_3 - 2z_4 + z_5) + 0.75(-z_3 + z_5) - 0.5z_4 = 0.5(1.5^2)$

Node numbers

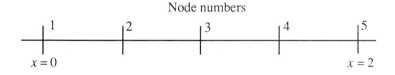

Fig. 6.4.1. Node numbering used in the solution of (6.4.1).

The problem boundary conditions are $x = 0$, $z = 1$ and $x = 2$, $z = 2$. Thus $z_1 = 1$ and $z_5 = 2$. Using these values, the above equations can be simplified and written in matrix form thus:

$$\begin{bmatrix} -44 & 22 & 0 \\ 28 & -68 & 36 \\ 0 & 46 & -108 \end{bmatrix} \begin{bmatrix} z_2 \\ z_3 \\ z_4 \end{bmatrix} = \begin{bmatrix} -17 \\ 4 \\ -107 \end{bmatrix}$$

This equation system can easily be solved using MATLAB as follows:

```
»a=[-44 22 0;28 -68 36;0 46 -108];
»b=[-17;4;-107];
»y=a\b

y =
    0.9357
    1.0987
    1.4587
```

Note that the rows in the above matrix equation can always be scaled in order to make the coefficient matrix symmetrical. This is important in a large problem.

In order to increase the accuracy of the solution we must increase the number of nodal points in order to decrease h. However, formulating the finite difference approximation by hand for a large number of nodes is a tedious and error prone process. The MATLAB function **twopoint** implements the process of solving the second-order boundary problem comprising differential equation (6.4.7) together with appropriate boundary conditions.

$$C(x)\frac{\mathrm{d}^2 z}{\mathrm{d}x^2} + D(x)\frac{\mathrm{d}z}{\mathrm{d}x} + E(x)z = F(x) \qquad (6.4.7)$$

The user must supply a vector listing the values of nodal points chosen. These do not have to be equispaced. The user must also supply vectors listing the values of $C(x)$, $D(x)$, $E(x)$ and $F(x)$ for the nodal points. Finally the user must provide the boundary conditions which can be in terms of either z or $\mathrm{d}z/\mathrm{d}x$.

```
function y=twopoint(x,C,D,E,F,flag1,flag2,p1,p2)
n=length(x)-1;
h(2:n+1)=x(2:n+1)-x(1:n);
h(1)=h(2); h(n+2)=h(n+1);
r(1:n+1)=h(2:n+2)./h(1:n+1);
s=1+r;
%flag1=1, y  specified at node 1.
%flag1=0, y' specified at node 1.
%flag2=1, y  specified at node n+1.
%flag2=0, y' specified at node n+1.
```

[Script continues...

```
if flag1==1
  y(1)=p1;
else
  slope0=p1;
end
if flag2==1
  y(n+1)=p2;
else
  slopen=p2;
end
W=zeros(n+1,n+1);
if flag1==1
  c0=3;
  W(2,2)=E(2)-2*C(2)/(h(2)^2*r(2));
  W(2,3)=2*C(2)/(h(2)^2*r(2)*s(2))+D(2)/(h(2)*s(2));
  b(2)=F(2)-y(1)*(2*C(2)/(h(2)^2*s(2))-D(2)/(h(2)*s(2)));
else
  c0=2;
  W(1,1)=E(1)-2*C(1)/(h(1)^2*r(1));
  W(1,2)=2*C(1)*(1+1/r(1))/(h(1)^2*s(1));
  b(1)=F(1)+slope0*(2*C(1)/h(1)-D(1));
end
if flag2==1
  c1=n-1;
  W(n,n)=E(n)-2*C(n)/(h(n)^2*r(n));
  W(n,n-1)=2*C(n)/(h(n)^2*s(n))-D(n)/(h(n)*s(n));
  b(n)=F(n)-y(n+1)*(2*C(n)/(h(n)^2*s(n))+D(n)/(h(n)*s(n)));
else
  c1=n;
  W(n+1,n+1)=E(n+1)-2*C(n+1)/(h(n+1)^2*r(n+1));
  W(n+1,n)=2*C(n+1)*(1+1/r(n+1))/(h(n+1)^2*s(n+1));
  b(n+1)=F(n+1)-slopen*(2*C(n+1)/h(n+1)+D(n+1));
end
for i=c0:c1
  W(i,i)=E(i)-2*C(i)/(h(i)^2*r(i));
  W(i,i-1)=2*C(i)/(h(i)^2*s(i))-D(i)/(h(i)*s(i));
  W(i,i+1)=2*C(i)/(h(i)^2*r(i)*s(i))+D(i)/(h(i)*s(i));
  b(i)=F(i);
end
z=W(flag1+1:n+1-flag2,flag1+1:n+1-flag2)\b(flag1+1:n+1-flag2)';
if flag1==1 & flag2==1, y=[y(1); z; y(n+1)]; end
if flag1==1 & flag2==0, y=[y(1); z]; end
if flag1==0 & flag2==1, y=[z; y(n+1)]; end
if flag1==0 & flag2==0, y=z; end
```

We can use this function to solve (6.4.1) for nine nodes using the following script:

```
x=0:.2:2; C=1+x.^2; D=x; E=-ones(1,11); F=x.^2;
flag1=1; p1=1; flag2=1; p2=2;
z=twopoint(x,C,D,E,F,flag1,flag2,p1,p2);
B=1/3; A=-sqrt(5)*B/2;
xx=0:.01:2;
zz=A*xx+B*sqrt(1+xx.^2)+B*(2+xx.^2);
plot(x,z,'o',xx,zz)
xlabel('x'); ylabel('z')
```

This script outputs the graph of Fig. 6.4.2. It is seen that the results from the finite difference analysis are very accurate. This is because the solution of the boundary problem, given by (6.4.2), is well approximated by a low-order polynomial.

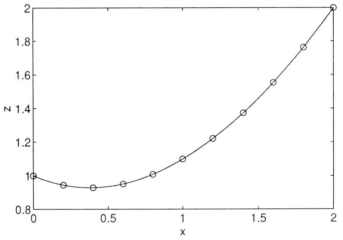

Fig. 6.4.2. Finite difference solution of $(1 + x^2)(d^2z/dx^2) + xdz/dx - z = x^2$.

Example 2. Determine the approximate solution of (6.4.3) subject to the boundary conditions that $z = 0$ at $x = 1$ and $dz/dx = 0$ at $x = e$. The exact eigensolutions are given by $\lambda_n = \{(2n + 1)\pi/2\}^2$ and $z_n = \sin\{(2n + 1)(\pi/2)\log_e|x|\}$, where $n = 0, 1, ..., \infty$. We will use the node numbering scheme shown in Fig. 6.4.3. To apply the boundary condition at $x = e$ we must consider the finite difference approximation for Dz at node 5 (i.e. at $x = e$) and make $Dz_5 = 0$. Applying (6.3.1) we have

Node numbers

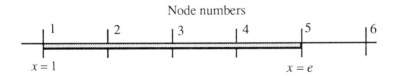

Fig. 6.4.3. Node numbering used in the solution of (6.4.3).

$$2hDz_5 = -z_4 + z_6 = 0 \qquad (6.4.8)$$

Note that we have been forced to introduce a fictitious node, node 6. However, from (6.4.8), $z_6 = z_4$.

Multiplying (6.4.3) by $2h^2$ gives

$$2x(h^2D^2z) + h(2hDz) = -\lambda 2x^{-1}h^2z$$

Thus

$$2x_i(z_{i-1} - 2z_i + z_{i+1}) + h(-z_{i-1} + z_{i+1}) = -\lambda 2x_i^{-1}h^2z_i$$

Now $L = e - 1 = 1.7183$. Thus $h = L/4 = 0.4296$. Applying (6.4.5) to nodes $2 - 5$ we have

At node 2: $2(1.4296)(z_1 - 2z_2 + z_3) + 0.4296(-z_1 + z_3) = -\lambda 2(1.4296)^{-1}(0.4296)^2 z_2$

At node 3: $2(1.8591)(z_2 - 2z_3 + z_4) + 0.4296(-z_2 + z_4) = -\lambda 2(1.8591)^{-1}(0.4296)^2 z_3$

At node 4: $2(2.2887)(z_3 - 2z_4 + z_5) + 0.4296(-z_3 + z_5) = -\lambda 2(2.2887)^{-1}(0.4296)^2 z_4$

At node 5: $2(2.7183)(z_4 - 2z_5 + z_6) + 0.4296(-z_4 + z_6) = -\lambda 2(2.7183)^{-1}(0.4296)^2 z_5$

Letting $z_1 = 0$ and $z_6 = z_4$ leads to

$$
\begin{bmatrix}
-5.7184 & 3.2887 & 0 & 0 \\
3.2887 & -7.4364 & 4.1478 & 0 \\
0 & 4.1478 & -9.1548 & 5.0070 \\
0 & 0 & 10.8731 & -10.8731
\end{bmatrix}
\begin{bmatrix} z_2 \\ z_3 \\ z_4 \\ z_5 \end{bmatrix}
$$

$$
= \lambda
\begin{bmatrix}
-0.2582 & 0 & 0 & 0 \\
0 & -0.1985 & 0 & 0 \\
0 & 0 & -0.1613 & 0 \\
0 & 0 & 0 & -0.1358
\end{bmatrix}
\begin{bmatrix} z_2 \\ z_3 \\ z_4 \\ z_5 \end{bmatrix}
$$

We can solve these equations using MATLAB as follows:

```
»a=[-5.7184 3.2887 0 0;3.2887 -7.4364 4.2478 0; ...
    0 4.1478 -9.1548 5.0070;0 0 10.8731 -10.8731];
»b=[-0.2582 0 0 0;0 -0.1985 0 0;0 0 -0.1613 0;0 0 0 -0.1358];
»[u,lambda]=eig(a,b)

u =
    0.3168   -0.6656    0.3025   -0.0152
    0.4931   -0.0912   -0.6954    0.1198
    0.5644    0.4426    0.2195   -0.4630
    0.5813    0.5939    0.6138    0.8781
```

[*Output continues...*

```
lambda =
    2.3228        0         0          0
         0  20.4014         0          0
         0        0   51.4264          0
         0        0         0   122.2829
```

The exact values for the lowest four eigenvalues are 2.4674, 22.2066, 61.6850 and 120.9027. The graph of Fig. 6.4.3 shows the first two eigenfunctions and the estimates derived from the array u above. Note that the values of u have been scaled to make the values of u corresponding to z_5 either 1 or –1.

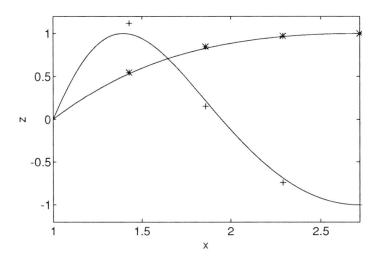

Fig. 6.4.3. The finite difference estimates for the first (*) and second (+) eigenfunctions of $x(d^2z/dx^2) + dz/dx + \lambda z/x = 0$. Solid lines show the exact eigenfunctions.

6.5 PARABOLIC PARTIAL DIFFERENTIAL EQUATIONS

The general second-order partial differential equation in terms of the independent variables x and y is given by (6.1.2). The equation is repeated here, except that y has been replaced by t.

$$A(x, t)\frac{\partial^2 z}{\partial x^2} + B(x, t)\frac{\partial^2 z}{\partial x\, \partial t} + C(x, t)\frac{\partial^2 z}{\partial t^2} + f\left(x, t, z, \frac{\partial z}{\partial x}, \frac{\partial z}{\partial t}\right) = 0 \qquad (6.5.1)$$

This equation will be a parabolic equation if $B^2 - 4AC = 0$. Parabolic equations are not defined in a closed domain but propagate in an open domain. For example, the one-dimensional heat-flow equation, which describes heat flow assuming no energy generation, is

$$K \frac{\partial^2 u}{\partial x^2} = \frac{\partial u}{\partial t}, \quad 0 < x < L \text{ and } t > 0 \tag{6.5.2}$$

where K is the thermal diffusivity and u is the temperature of the material. Comparing (6.5.2) with (6.5.1), we see that A, B and C of (6.5.1) are K, 0 and 0 respectively so that the term $B^2 - 4AC$ of (6.5.1) is zero and the equation is parabolic.

In order to solve this equation, boundary conditions must be specified at $x = 0$ and $x = L$ and initial conditions when $t = 0$ must also be given. To develop a finite difference solution we divide the spatial domain into n sections, each of length h, so that $h = L/n$, and consider as many time steps as required, each of duration k. A finite difference approximation for (6.5.2) at node (i, j) is obtained by replacing $\partial^2 u/\partial x^2$ by the central difference approximation (6.3.2) and $\partial u/\partial t$ by the forward difference approximation (6.3.5) to give

$$K \left(\frac{u_{i-1,j} - 2u_{i,j} + u_{i+1,j}}{h^2} \right) = \left(\frac{-u_{i,j} + u_{i,j+1}}{k} \right) \tag{6.5.3}$$

or

$$u_{i,j+1} = u_{i,j} + \alpha(u_{i-1,j} - 2u_{i,j} + u_{i+1,j}), \quad i = 0, 1, ..., n; \quad j = 0, 1, ... \tag{6.5.4}$$

In (6.5.4) $\alpha = Kk/h^2$. Node (i, j) is the point $x = ih$ and at time jk. Equation (6.5.4) allows us to determine $u_{i,j+1}$, i.e. u at time $j + 1$ from values of u at time j. Values of $u_{i,0}$ are provided by the initial conditions, values of $u_{0,j}$ and $u_{n,j}$ are obtained from the boundary conditions. This method of solution is called the explicit method.

In the numeric solution of parabolic partial differential equations, solution stability and convergence are important. It can be proved that when using the explicit method we must make $\alpha \leq 0.5$ to ensure a steady decay of the entire solution. This requirement means that the grid separation in time must sometimes be very small, necessitating a very large number of time steps.

An alternative finite difference approximation for (6.5.2) is obtained by considering node $(i, j+1)$. We again approximate $\partial^2 u/\partial x^2$ by the central difference approximation (6.3.2) but we approximate $\partial u/\partial t$ by the backward difference approximation (6.3.6) to give

$$K \left(\frac{u_{i-1,j+1} - 2u_{i,j+1} + u_{i+1,j+1}}{h^2} \right) = \left(\frac{-u_{i,j} + u_{i,j+1}}{k} \right) \tag{6.5.5}$$

This equation is identical to (6.5.3) except that approximation is made at the $(j + 1)$th time step instead of at the jth time step. Rearranging (6.5.5) with $\alpha = Kk/h^2$ gives

$$(1 + 2\alpha)u_{i,j+1} - \alpha(u_{i+1,j+1} + u_{i-1,j+1}) = u_{i,j} \tag{6.5.6}$$

where $i = 0, 1, ..., n; \ j = 0, 1, ...$. The three variables on the left-hand side of this equation are unknown. However, if we have a grid of $n + 1$ spatial points, then at time $j + 1$ there are $n - 1$ unknown nodal values and two known boundary values. We can assemble the set of $n - 1$ equations of the form of (6.5.6) thus:

$$\begin{bmatrix} (1+2\alpha) & -\alpha & 0 & \cdots & 0 \\ -\alpha & (1+2\alpha) & -\alpha & \cdots & 0 \\ 0 & -\alpha & (1+2\alpha) & \cdots & 0 \\ \vdots & \vdots & \vdots & & \vdots \\ 0 & 0 & 0 & \cdots & -\alpha \\ 0 & 0 & 0 & \cdots & (1+2\alpha) \end{bmatrix} \begin{bmatrix} u_{1,j+1} \\ u_{2,j+1} \\ u_{3,j+1} \\ \vdots \\ u_{n-2,j+1} \\ u_{n-1,j+1} \end{bmatrix} = \begin{bmatrix} u_{1,j} + \alpha u_0 \\ u_{2,j} \\ u_{3,j} \\ \vdots \\ u_{n-2,j} \\ u_{n-1,j} + \alpha u_n \end{bmatrix}$$

Note that u_0 and u_n are the known boundary conditions, assumed to be independent of time. By solving the equation system above we determine $u_1, u_2, ..., u_{n-1}$ at time step $j + 1$ from $u_1, u_2, ..., u_{n-1}$ at time step j. This approach is called the implicit method. Compared with the explicit method each time step requires more computation but the method has the significant advantage that it is unconditionally stable. However, although stability does not place any restriction on α, h and k must be chosen to keep the discretisation error small to maintain accuracy.

The function heat listed below implements an implicit finite difference solution for the parabolic differential equation (6.5.2).

```
function u=heat(nx,hx,nt,ht,init,lowb,hib,K)
% Parabolic equ'n - ht flow. Implicit method.
% nx, hx are number and size of x panels, index i
% nt, ht are number and size of t panels, index j
% init is a vector of nx+1 initial values of function.
% lowb & hib are boundaries at low and hi values of x.
alpha=K*ht/hx^2;
A=zeros(nx-1,nx-1); u=zeros(nt+1,nx+1);
u(:,1)=lowb*ones(nt+1,1);
u(:,nx+1)=hib*ones(nt+1,1);
u(1,:)=init;
A(1,1)=1+2*alpha; A(1,2)=-alpha;
for i=2:nx-2
  A(i,i)=1+2*alpha;
  A(i,i-1)=-alpha; A(i,i+1)=-alpha;
end
A(nx-1,nx-2)=-alpha; A(nx-1,nx-1)=1+2*alpha;
b(1,1)=init(2)+init(1)*alpha;
for i=2:nx-2, b(i,1)=init(i+1); end
b(nx-1,1)=init(nx)+init(nx+1)*alpha;
[L,U]=lu(A);
for j=2:nt+1
  y=L\b; x=U\y;
  u(j,2:nx)=x'; b=x;
  b(1,1)=b(1,1)+lowb*alpha;
  b(nx-1,1)=b(nx-1,1)+hib*alpha;
end
```

We now use the function heat to study how the temperature distribution in a brick wall varies with time. The wall is 0.3 m thick and is initially at a uniform temperature of 100 °C. For brickwork, $K = 5 \times 10^{-7}$ m²/s. If the temperature of both surfaces is suddenly lowered to 20 °C and kept at this temperature, we wish to plot the subsequent variation of temperature through the wall at 7.33 min (440 s) intervals for 366.67 min (22000 s).

To study this problem we will use a mesh with 15 subdivisions of x and 50 subdivisions of t. This corresponds to $\alpha = 0.55$.

```
K=5e-7; hx=0.02; nx=15; % hx*nx=0.3
ht=440; nt=50;          % ht*nt=22000
init=100*ones(1,nx+1);
lowb=20; hib=20;
u=heat(nx,hx,nt,ht,init,lowb,hib,K);
surfl(u)
axis([0 16 0 50 0 120])
view([-217 30])
xlabel('x - node nos.'); ylabel('time - node nos.');
zlabel('temperature')
```

Running this script gives the plot shown in Fig. 6.5.1. This plot shows how the temperature across the wall decreases with time. Fig. 6.5.2 shows the variation of temperature with time at the centre of the wall, calculated by both the implicit method (using the MATLAB function heat) and the explicit method. In the latter case a MATLAB function is not provided. It is seen that the solution determined using the explicit method becomes unstable with increasing time. We would expect this because the mesh size has been chosen to make $\alpha = 0.55$.

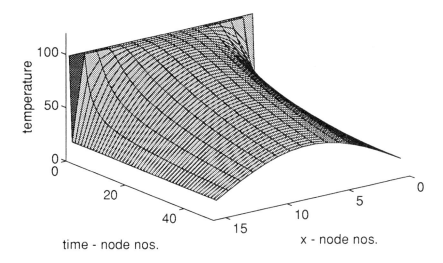

Fig. 6.5.1. Distribution of temperature through a wall.

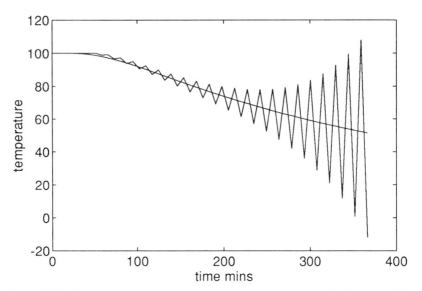

Fig. 6.5.2. Variation of temperature in the centre of a wall. The steadily
decaying solution was generated using the implicit method;
the oscillating solution was generated using the explicit method.

6.6 HYPERBOLIC PARTIAL DIFFERENTIAL EQUATIONS

Consider the following equation:

$$c^2 \frac{\partial^2 u}{\partial x^2} = \frac{\partial^2 u}{\partial t^2}, \quad 0 < x < L \text{ and } t > 0 \tag{6.6.1}$$

This is the one-dimensional wave equation, and like the heat-flow problem of section
6.5, its solution usually propagates in an open domain. Equation (6.6.1) describes the
wave in a taught string where c is the velocity of propagation of the waves in the string.
Comparing (6.6.1) with (6.5.1) we see that $B^2 - 4AC = -4c^2(-1)$. Since c^2 must be
positive $B^2 - 4AC > 0$ and the equation is hyperbolic. Equation (6.6.1) is subject to
boundary conditions when $x = 0$ and $x = L$ and also subject to initial conditions when
$t = 0$.

We now develop equivalent finite difference approximations for these equations.
We divide L into n sections so that $h = L/n$ and consider time steps of duration k.
Approximating (6.6.1) by central finite difference approximations based on (6.3.2) at
node (i, j) we have

$$c^2 \left(\frac{u_{i-1,j} - 2u_{i,j} + u_{i+1,j}}{h^2} \right) = \left(\frac{u_{i,j-1} - 2u_{i,j} + u_{i,j+1}}{k^2} \right)$$

or

$$(u_{i-1,j} - 2u_{i,j} + u_{i+1,j}) - (1/\alpha^2)(u_{i,j-1} - 2u_{i,j} + u_{i,j+1}) = 0$$

where $\alpha^2 = c^2 k^2 / h^2$, $i = 0, 1, ..., n$ and $j = 0, 1, ...$. Node (i, j) is the point $x = ih$ at time $t = jk$. Rearranging the above equation gives

$$u_{i,j+1} = \alpha^2(u_{i-1,j} + u_{i+1,j}) + 2(1 - \alpha^2)u_{i,j} - u_{i,j-1} \qquad (6.6.2)$$

When $j = 0$ equation (6.6.2) becomes

$$u_{i,1} = \alpha^2(u_{i-1,0} + u_{i+1,0}) + 2(1 - \alpha^2)u_{i,0} - u_{i, -1} \qquad (6.6.3)$$

To solve a hyperbolic partial differential equation, initial values of $u(x)$ and $\partial u/\partial t$ must be specified. Let these values be U_i and V_i respectively, where $i = 0, 1, ..., n$. We can replace $\partial u/\partial t$ by its central finite difference approximation based on (6.3.1) thus:

$$V_i = (-u_{i, -1} + u_{i,1})/(2k)$$

Hence

$$-u_{i, -1} = 2kV_i - u_{i,1} \qquad (6.6.4)$$

In (6.6.3) we replace $u_{i,0}$ by U_i and $u_{i, -1}$ by using (6.6.4) to give

$$u_{i,1} = \alpha^2(U_{i-1} + U_{i+1}) + 2(1 - \alpha^2)U_i + 2kV_i - u_{i,1}$$

so that

$$u_{i,1} = \alpha^2(U_{i-1} + U_{i+1})/2 + (1 - \alpha^2)U_i + kV_i \qquad (6.6.5)$$

Equation (6.6.5) is the starting equation and allows us to determine the values of u at time step $j = 1$. Once we obtain these values we can use (6.6.2). Equations (6.6.5) followed by (6.6.2) provide an explicit method of solution. In order to ensure stability, the parameter α should be equal to or less than one. However, if α is less than one the solution becomes less accurate.

The following function fwave implements an explicit finite difference solution for (6.6.1).

```
function u=fwave(nx,hx,nt,ht,init,initslope,lowb,hib,c)
% nx, hx are number and size of x panels, index i
% nt, ht are number and size of t panels, index j
alpha=c*ht/hx
u=zeros(nt+1,nx+1);
u(:,1)=lowb; u(:,nx+1)=hib; u(1,:)=init;
for i=2:nx
  u(2,i)=alpha^2*(init(i+1)+init(i-1))/2+(1-alpha^2)*init(i) ...
  +ht*initslope(i);
end
```

[Script continues...

```
for j=2:nt
  for i=2:nx
    u(j+1,i)=alpha^2*(u(j,i+1)+u(j,i-1))+(2-2*alpha^2)*u(j,i) ...
    -u(j-1,i);
  end
end
```

We now use the function fwave to examine the effect of displacing the boundary at one end of a taught string by 10 units in a positive direction for the time period $t = 0.1$ to $t = 4$ units.

```
hx=1/10; ht=1/10; nx=16; nt=40; c=1;t=0:nt;
hib=zeros(nt+1,1); lowb=zeros(nt+1,1);
lowb(2,1)=10; lowb(3,1)=10; lowb(4,1)=10; lowb(5,1)=10;
init=zeros(1,nx+1); initslope=zeros(1,nx+1);
u=fwave(nx,hx,nt,ht,init,initslope,lowb,hib,c);
surfl(u)
axis([0 16 0 40 -10 10])
xlabel('position along string'); ylabel('time');
zlabel('vertical displacement')
```

Running this script produces the following output, together with Fig. 6.6.1.

```
alpha =
    1
```

Fig. 6.6.1 shows that the disturbance at the boundary travels along the string. At the other boundary it is reflected and becomes a negative disturbance. This process of reflection and reversal continues at each boundary. The disturbance travels at a velocity

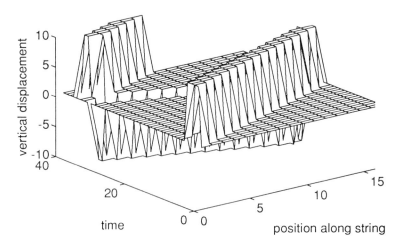

Fig. 6.6.1. Solution of (6.6.1) subject to specific
boundary and initial conditions.

c and its shape does not change. Similarly pressure fluctuations do not change as they travel along a speaking tube and if pressure fluctuations representing the sound "HELLO" enter the tube, the sound "HELLO" is detected at the other end.

6.7 ELLIPTIC PARTIAL DIFFERENTIAL EQUATIONS

The solution of a second-order elliptic partial differential equation is determined over a closed region and the shape of the boundary and its condition at every point must be specified. Some important second-order elliptic partial differential equations, which arise naturally in the description of physical systems, are

Laplace's equation:	$\nabla^2 z = 0$	(6.7.1)
Poisson's equation:	$\nabla^2 z = F(x,y)$	(6.7.2)
Helmholtz's equation:	$\nabla^2 z + G(x,y)z = F(x,y)$	(6.7.3)

where $\nabla^2 z = \partial^2 z/\partial x^2 + \partial^2 z/\partial y^2$ and $z(x,y)$ is an unknown function. Note that the Laplace and Poisson equations are special cases of Helmholtz's equation. In general these equations must satisfy boundary conditions that are specified in terms of either the function value or the derivative of the function normal to the boundary. Furthermore a problem can have mixed boundary conditions.

If we compare (6.7.1) to (6.7.3) with the standard second-order partial differential equation in two variables, i.e.

$$A(x,y)\frac{\partial^2 z}{\partial x^2} + B(x,y)\frac{\partial^2 z}{\partial x\,\partial y} + C(x,y)\frac{\partial^2 z}{\partial y^2} + f\left(x,y,z,\frac{\partial z}{\partial x},\frac{\partial z}{\partial y}\right) = 0$$

we see that in each case $A = C = 1$ and $B = 0$, so that $B^2 - 4AC < 0$, confirming that the equations are elliptic.

The Laplace equation is homogeneous and if a problem has boundary conditions that are also homogeneous then the solution, $z = 0$, will be trivial. Similarly in (6.7.3), if $F(x,y) = 0$ and the problem boundary conditions are homogeneous then $z = 0$. However, in (6.7.3) we can scale $G(x,y)$ by a factor λ, so that (6.7.3) becomes

$$\nabla^2 z + \lambda G(x,y)z = 0 \tag{6.7.4}$$

This is a characteristic or eigenvalue problem and we can determine values of λ and corresponding non-trivial values of $z(x,y)$.

The elliptic equations (6.7.1) to (6.7.4) can only be solved in a closed form for a limited number of situations. For most problems it is necessary to use a numerical approximation. Finite difference methods are relatively simple to apply, particularly for rectangular regions. We will now use the finite difference approximation for $\nabla^2 z$, (6.3.7) or (6.3.8), to the solve some elliptic partial differential equations over a rectangular domain.

Example 1. Laplace's equation. Determine the distribution of temperature in a rectangular plane section, subject to a temperature distribution around its edges as follows:

$$x = 0, T = 100y; \quad x = 3, T = 250y; \quad y = 0, T = 0 \text{ and } y = 2, T = 200 + (100/3)x^2$$

The section, the boundary temperature distribution and two chosen nodes are shown in Fig. 6.7.1.

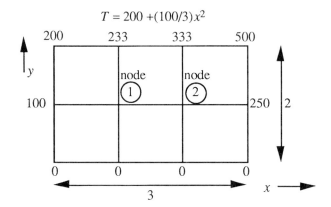

Fig. 6.7.1. Temperature distribution around a plane section.
Location of nodes 1 and 2 is shown.

The temperature distribution is described by Laplace's equation. Solving this equation by the finite difference method, we apply (6.3.8) to nodes 1 and 2 of the mesh shown in Fig. 6.7.1. This gives

$$(233.33 + T_2 + 0 + 100 - 4T_1)/h^2 = 0$$
$$(333.33 + 250 + 0 + T_1 - 4T_2)/h^2 = 0$$

where T_1 and T_2 are the unknown temperatures at nodes 1 and 2 respectively and $h = 1$. Rearranging these equations gives

$$\begin{bmatrix} -4 & 1 \\ 1 & -4 \end{bmatrix} \begin{bmatrix} T_1 \\ T_2 \end{bmatrix} = \begin{bmatrix} -333.33 \\ -583.33 \end{bmatrix}$$

Solving this equation we have $T_1 = 127.78$ and $T_2 = 177.78$.

If we require a more accurate solution of Laplace's equation then we must use more nodes and the computation burden increases rapidly. The MATLAB function ellipgen listed below uses the finite difference approximation (6.3.7) to solve the general elliptic partial differential equations (6.7.1) to (6.7.4) for a rectangular domain only. The function is also limited to problems in which the boundary value is specified by values of the function $z(x,y)$, not its derivative. If the user calls the function with 10 arguments the function solves (6.7.1) to (6.7.3), see Examples 1 and 2 below. Calling it with six arguments causes it to solve (6.7.4), see Example 3 below.

```
function [a,om]=ellipgen(nx,hx,ny,hy,G,F,bx0,bxn,by0,byn)
% (∂2/∂x2+∂2/∂y2)*z+G(x,y)*z=F(x,y) over rectangular region.
% hx, hy are panel sizes in x and y directions,
% nx, ny are number of panels in x and y directions.
% F and G are (ny+1,nx+1) arrays representing F(x,y), G(x,y).
% bx0 and bxn are vectors of boundary conditions at x0 and xn.
% each beginning at y0. Each is (ny+1) elements.
% by0 and byn are vectors of boundary conditions at y0 and yn.
% each beginning at x0. Each is (nx+1) elements.
% If the last 4 parameters (boundary conditions) are omitted,
% function solves (∂2/∂x2+∂2/∂y2)*z+lambda*G(x,y)*z=0
% over rectangular region. F is then a scalar and specifies
% the eigenvector to be returned in array a.
% The vector om lists the eigenvalues lambda.
nmax=(nx-1)*(ny-1); r=hy/hx;
a=zeros(ny+1,nx+1); p=zeros(ny+1,nx+1);
if nargin==6
  case=0;
  mode=F;
end
if nargin==10
  test=0;
  if F==zeros(ny+1,nx+1), test=1; end
  if bx0==zeros(1,ny+1), test=test+1; end
  if bxn==zeros(1,ny+1), test=test+1; end
  if by0==zeros(1,nx+1), test=test+1; end
  if byn==zeros(1,nx+1), test=test+1; end
  if test==5
    disp('WARNING - problem has trivial solution, z = 0.')
    disp('To obtain eigensolution use 6 parameters only.')
    break
  end
  bx0=bx0(1,ny+1:-1:1); bxn=bxn(1,ny+1:-1:1);
  a(1,:)=byn; a(ny+1,:)=by0;
  a(:,1)=bx0'; a(:,nx+1)=bxn';
  case=1;
end
for i=2:ny
  for j=2:nx
    nn=(i-2)*(nx-1)+(j-1);
    q(nn,1)=i; q(nn,2)=j; p(i,j)=nn;
  end
end
C=zeros(nmax,nmax); e=zeros(nmax,1); om=zeros(nmax,1);
if case==1, g=zeros(nmax,1); end
```

[Script continues...

```
for i=2:ny
  for j=2:nx
    nn=p(i,j); C(nn,nn)=-(2+2*r^2); e(nn)=hy^2*G(i,j);
    if case==1, g(nn)=g(nn)+hy^2*F(i,j); end
    if p(i+1,j)~=0
      np=p(i+1,j); C(nn,np)=1;
    else
      if case==1, g(nn)=g(nn)-by0(j); end
    end
    if p(i-1,j)~=0
      np=p(i-1,j); C(nn,np)=1;
    else
      if case==1, g(nn)=g(nn)-byn(j); end
    end
    if p(i,j+1)~=0
      np=p(i,j+1); C(nn,np)=r^2;
    else
      if case==1, g(nn)=g(nn)-r^2*bxn(i); end
    end
    if p(i,j-1)~=0
      np=p(i,j-1); C(nn,np)=r^2;
    else
      if case==1, g(nn)=g(nn)-r^2*bx0(i); end
    end
  end
end
if case==1
  C=C+diag(e);
  z=C\g;
  for nn=1:nmax
    i=q(nn,1); j=q(nn,2);
    a(i,j)=z(nn);
  end
else
  [u,lam]=eig(C,-diag(e));
  [om,k]=sort(diag(lam));
  u=u(:,k);
  for nn=1:nmax
    i=q(nn,1); j=q(nn,2);
    a(i,j)=u(nn,mode);
  end
end
```

We now give examples of the application of this function.

Example 1. (Continued.) Use function ellipgen to solve Laplace's equation over a rectangular region subject to the boundary conditions shown in Fig. 6.7.1. The following script calls the function to solve this problem using a 6 x 6 mesh.

```
nx=6; ny=6; hx=0.5; hy=0.3333;
by0=[0 0 0 0 0 0 0];
byn=[200 208.33 233.33 275 333.33 408.33 500];
bx0=[0 33.33  66.67 100 133.33 166.67 200];
bxn=[0 83.33 166.67 250 333.33 416.67 500];
F=zeros(ny+1,nx+1); G=F;
a=ellipgen(nx,hx,ny,hy,G,F,bx0,bxn,by0,byn);
aa=flipud(a);
contour(aa)
xlabel('node numbers in x direction');
ylabel('node numbers in y direction');
```

The output from this script is the contour plot shown in Fig. 6.7.2. The temperature is not shown on the contour plot; if required it can be obtained from *aa*.

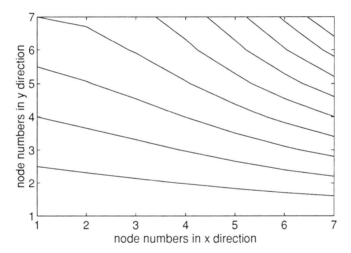

Fig. 6.7.2. Finite difference estimate for the temperature distribution for the problem defined in Fig. 6.7.1.

Example 2. Poisson's equation. Determine the deflection of a uniform square membrane, held at its edges and subject to a distributed load which can be approximated to a unit load at each node. This problem is described by Poisson's equation, (6.7.2), where $F(x,y)$ specifies the load on the membrane. We use the following script to determine the deflection of this membrane using the MATLAB function ellipgen.

```
nx=6; ny=6; hx=1/6; hy=1/6;
by0=[0 0 0 0 0 0 0];   byn=[0 0 0 0 0 0 0];
bx0=[0 0 0 0 0 0 0];   bxn=[0 0 0 0 0 0 0];
F=-ones(ny+1,nx+1); G=zeros(nx+1,ny+1);
a=ellipgen(nx,hx,ny,hy,G,F,bx0,bxn,by0,byn);
surfl(a)
axis([1 7 1 7 0 0.1])
xlabel('x-node nos.');ylabel('y-node nos.');zlabel('displacement');
```

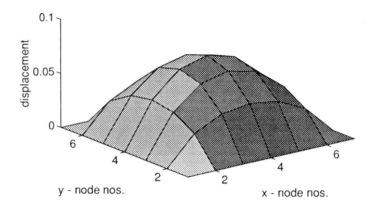

Fig. 6.7.3. Deflection of a square membrane subject to a distributed load.

Running this script gives the output shown in Fig. 6.7.3. The displacement at the centre of the membrane is 0.0721 by this analysis. This compares with the exact value of 0.0737.

Example 3. Characteristic value problem. Determine the natural frequencies and mode shapes of a freely vibrating square membrane held at its edges. This problem is described by the eigenvalue problem (6.7.4). The natural frequencies are related to the eigenvalues and the mode shapes are the eigenvectors. The MATLAB script below determines the eigenvalues and vectors. It calls the function ellipgen and outputs a list of eigenvalues and provides Fig. 6.7.4 showing the second mode shape of the membrane.

```
nx=6; ny=9; hx=1/6; hy=1/6; G=ones(10,7); mode=2;
[a,om]=ellipgen(nx,hx,ny,hy,G,mode);
om(1:5)
mesh(a)
axis([1 8 1 10 -0.5 0.5])
xlabel('y - node nos.'); ylabel('x - node nos.');
zlabel('relative displacement');
```

Running this script gives

```
ans =
    13.9883
    26.4910
    40.3421
    45.6462
    52.8448
```

These eigenvalues compare with the exact values given in Table 6.7.1.

Table 6.7.1.

FD approx.	Exact	Error, %
13.9883	14.2561	1.88
26.4910	27.4156	3.37
40.3421	43.8649	8.03
45.6462	49.3480	7.50
52.8448	57.0244	7.33

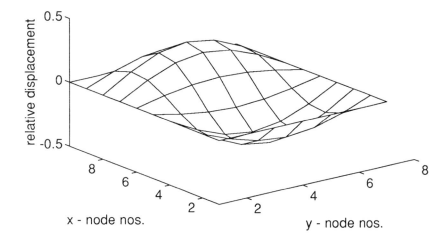

Fig. 6.7.4. Finite difference approximation of the second mode
of vibration of a uniform membrane.

6.8 SUMMARY

We have examined the application of finite difference methods to a broad range of
second-order ordinary and partial differential equations. A major problem in the
development of scripts is the difficulty of accounting for the wide variety of boundary
conditions and boundary shapes that can occur. Software packages have been developed
to solve the partial differential equations that arise in computational fluid dynamics and
continuum mechanics using either finite difference or finite element methods but they
are both complex and expensive because they allow the user total freedom to define
boundary shapes and conditions.

PROBLEMS

6.1. Classify the following second-order partial differential equations:

(i) $\dfrac{\partial^2 y}{\partial t^2} + a\,\dfrac{\partial^2 y}{\partial x\,\partial t} + \dfrac{1}{4}(a^2 - 4)\dfrac{\partial^2 y}{\partial x^2} = 0$

(ii) $\dfrac{\partial u}{\partial t} - \dfrac{\partial}{\partial x}\!\left(A(x,t)\dfrac{\partial u}{\partial x}\right) = 0$

(iii) $\dfrac{\partial^2 \phi}{\partial x^2} = k\,\dfrac{\partial^2(\phi^2)}{\partial y^2}$ where $k > 0$

6.2. Use the shooting method to solve $y'' + y' - 6y = 0$, where the prime denotes differentiation with respect to x, given the boundary conditions $y(0) = 1$ and $y(1) = 2$. Note that an illustrative script for the shooting method is given in section 6.2. Use trial slopes in the range $-3:0.5:2$. Compare your results with those you obtain using the finite difference method with 10 divisions. The finite difference method is implemented by function twopoint. Note that the exact solution is

$$y = 0.2657\exp(2x) + 0.7343\exp(-3x)$$

6.3. (i) Use the shooting method to solve $y'' - 62y' + 120y = 0$ where the prime denotes differentiation with respect to x, given the boundary conditions $y(0) = 0$ and $y(1) = 2$. Solve this equation by applying the shooting method, using trial slopes in the range $-0.5:0.1:0.5$. Note that the exact solution is

$$y = 1.751302152539304 \times 10^{-26}\{\exp(60x) - \exp(2x)\}$$

(ii) By substituting $x = 1 - p$ in the original differential equation show that $y'' + 62y' + 120y = 0$ where the prime denotes differentiation with respect to p. Note that the boundary conditions of this problem are $y(0) = 2$ and $y(1) = 0$. Solve this equation by applying the shooting method, using trial slopes in the range 0 to -150 in steps of -30 at $p = 0$. Note that a very good approximation to the solution is $y = 2\exp(-60p)$.

Compare the two answers you obtain for (i) and (ii). Note that an illustrative script for the shooting method is given in section 6.2. Also solve (i) and (ii) using the finite difference method, implemented in twopoint. Use 10 divisions and repeat with 50 divisions. You should plot your answers and compare with a plot of the exact solution.

6.4. Solve the boundary value problem $xy'' + 2y' - xy = e^x$ given that $y(0) = 0.5$ and $y(2) = 3.694528$ using the finite difference method implemented by the function twopoint. Use 10 divisions in the finite difference solution and plot the results, together with the exact solution, $y = \exp(x)/2$.

6.5. Determine the finite difference equivalence of the characteristic value problem defined by $y'' + \lambda y = 0$ where $y(0) = 0$ and $y(2) = 0$. Use 20 divisions in the finite difference method. Then solve the finite difference equations using the MATLAB function eig to determine the dominant value of λ.

6.6. Solve the parabolic equation (6.5.2) with $K = 1$, subject to the following boundary conditions: $u(0, t) = 0$, $u(1, t) = 10$, $u(x, 0) = 0$ for all x except $x = 1$. When $x = 1$, $u(1, 0) = 10$. Use the function heat to determine the solution for $t = 0$ to 0.5 in steps of 0.01 with 20 divisions of x. You should plot the solution for ease of visualisation.

6.7. Solve the wave equation, (6.6.1) with $c = 1$, subject to the following boundary and initial conditions: $u(t, 0) = u(t, 1) = 0$, $u(0, x) = \sin(\pi x) + 2\sin(2\pi x)$ and $u_t(0, x) = 0$ where the subscript t denotes partial differentiation with respect to t. Use the function fwave to determine the solution for $t = 0$ to 4.5 in steps of 0.05 and use 20 divisions of x. Plot your results and compare with a plot of the exact solution, which is given by $u = \sin(\pi x)\cos(\pi t) + 2\sin(2\pi x)\cos(2\pi t)$.

6.8. Solve the equation

$$\nabla^2 V + 4\pi^2(x^2 + y^2)V = 4\pi \cos\{\pi(x^2 + y^2)\}$$

over the square region $0 \le x \le 0.5$ and $0 \le y \le 0.5$. The boundary conditions are

$$V(x, 0) = \sin(\pi x^2), \ V(x, 0.5) = \sin\{\pi(x^2 + 0.25)\}$$
$$V(0, y) = \sin(\pi y^2), \ V(0.5, y) = \sin\{\pi(y^2 + 0.25)\}$$

Use the function ellipgen to solve this equation with 15 divisions of x and y. Plot your results and compare with a plot of the exact solution, which is given by $V = \sin\{\pi(x^2 + y^2)\}$.

6.9. Solve the eigenvalue problem $\nabla^2 z + \lambda G(x, y)z = 0$ over a rectangular region bounded by $0 \le x \le 1$ and $0 \le y \le 1.5$, $z = 0$ at all boundaries. Use the function ellipgen with six divisions in both directions. The function $G(x, y)$ over this grid is given by the MATLAB statements G=ones(10,7); G(4:7,3:5)=3*ones(4,3); This represents a membrane with a central area thicker than its periphery. The eigenvalues are related to the natural frequencies of this membrane.

6.10. Solve Poisson's equation $\nabla^2\phi + 2 = 0$ over the region shown below with $a = 1$. At the boundary $\phi = 0$. You will have to assemble the finite difference equation by hand, applying (6.3.8) to the 10 nodes, and then use MATLAB to solve the resulting linear equation system.

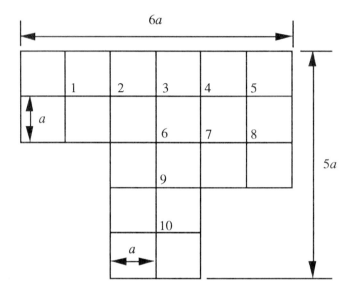

7

Fitting functions to data

7.1 INTRODUCTION

In this chapter we consider a variety of methods for fitting functions to data, describe some of the MATLAB toolbox functions that are available for this purpose and develop some additional ones. The application of these functions is illustrated by appropriate examples. We begin by examining polynomial interpolation.

7.2 INTERPOLATION USING POLYNOMIALS

Suppose y is some unknown function of x. Given a table of values of x and y, we may wish to obtain a value of y corresponding to a value of x that is not tabulated. Interpolation implies that the untabulated value of x is within the range of the tabulated data. If the untabulated x is outside this range the process is called *extrapolation* and is often less accurate.

The simplest form of interpolation is linear interpolation. In this method only the pair of data points enclosing the required value are used. Thus if (x_0, y_0) and (x_1, y_1) are two adjacent data points in a tabulation, to obtain the value of y corresponding to an x where $x_0 < x < x_1$ we fit the straight line $y = ax + b$ to these points and evaluate y thus:

$$y = \{y_0(x_1 - x) + y_1(x - x_0)\}/(x_1 - x_0) \tag{7.2.1}$$

We may use the MATLAB function `interp1` for this purpose. For example, consider the tabulation of the function $y = x^{1.9}$ for $x = 1, 2, ..., 5$. If we require estimates of y for $x = 2.5$ and 3.8 we may use `interp1` setting the third parameter as `'linear'` to obtain linear interpolation as follows:

```
»x=1:5;
»y=x.^1.9;
»interp1(x,y,[2.5,3.8],'linear')
ans =
    5.8979
   12.7558
```

The correct answers are $y = 5.7028$ and $y = 12.6354$ corresponding to $x = 2.5$ and 3.8 respectively.

Interpolation becomes more accurate when more of the tabulated data values are used because we can use a higher degree polynomial. A polynomial of degree n can be adjusted to pass through $n + 1$ data points. We do not need to know the coefficients of the polynomial explicitly but they are used implicitly in the procedure to estimate y for a given value of x. For example, MATLAB allows cubic interpolation by calling interp1 with the third parameter set as 'cubic'. The following example implements cubic interpolation using the same data as the previous example.

```
»interp1(x,y,[2.5,3.8],'cubic')
ans =
    5.7021
   12.6359
```

It can be seen that the cubic interpolation has given a much more accurate result.

An algorithm which provides an efficient method for fitting any degree polynomial to data is Aitken's algorithm. In this procedure a sequence of polynomial functions are fitted to the data. As the degree of the polynomial is increased more of the data points are used and the accuracy of the interpolation improves.

Aitken's algorithm proceeds as follows. Suppose we have five pairs of data values labelled 1, 2, ..., 5 and we wish to determine y^*, the value of y corresponding to a given x^*. Initially the algorithm determines straight lines (i.e. first degree polynomials) that pass through data points 1 and 2, 1 and 3, 1 and 4, and 1 and 5 as shown in Fig. 7.2.1(a). These four straight lines allow the procedure to determine four, probably poor, estimates for y^*.

Using $x_2, x_3, ..., x_5$ from the tabulated data and the four estimates of y^* the algorithm repeats the procedure above using these new points but this now provides second degree polynomials through the sets of data points $\{1, 2, 3\}$, $\{1, 2, 4\}$ and $\{1, 2, 5\}$ as shown in Fig. 7.2.1(b). From these second degree polynomials the procedure determines three improved estimates for y^*.

Using x_3, x_4, x_5 from the tabulated data and these three new estimates for y^*, the algorithm computes the third degree polynomials passing through the sets of data points $\{1, 2, 3, 4\}$ and $\{1, 2, 3, 5\}$ as shown in Fig. 7.2.1(c) to allow the procedure to determine two further improved estimates for y^*. Finally a fourth degree polynomial is computed that fits all the data. This fourth degree polynomial provides the best estimate for y^* as shown in Fig. 7.2.1(d).

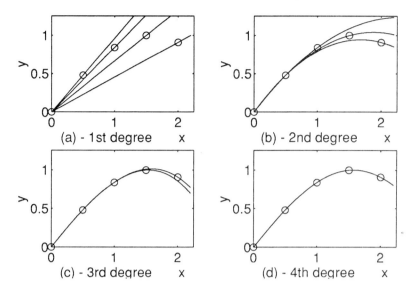

Fig. 7.2.1. Increasing the degree of the polynomial fit.

Aitken's algorithm has two advantages. First of all it is very efficient. Each new estimate for y^* requires only two multiplications and one division so that for $n + 1$ data points, the estimate using all the data requires $n(n + 1)$ multiplications and $n(n + 1)/2$ divisions. It is interesting to note that if we attempted to determine the coefficients of a polynomial passing through $n + 1$ data points by assembling a set of $n + 1$ linear equations, then in addition to the computation required to assemble the equations we would require $(n + 1)^3/2$ multiplications and divisions to solve them. The second advantage of Aitken's algorithm is that the process of fitting higher and higher degree polynomials to more and more of the data can be terminated when we note that the estimate of y^* is no longer changing significantly.

The MATLAB function aitken shown below implements Aitken's algorithm. The user must provide a set of data for the MATLAB vectors x and y. The function then determines a value of y corresponding to xval. The function provides the best value obtained and the Aitken table showing all the intermediate values obtained.

```
function Q=aitken(x,y,xval)
n=length(x); P=zeros(n);
P(1,:)=y;
for j=1:n-1
  for i=j+1:n
    P(j+1,i)=(P(j,i)*(xval-x(j))-P(j,j)*(xval-x(i)))/(x(i)-x(j));
  end
end
Q=P(n,n);
for i=1:n
  disp([x(i)  P(1:i,i)']);
end
```

We will now use this function to determine the reciprocal of 1.03 from a table of 10 equispaced values of x in the range 1 to 2 and $y = 1/x$. The script below calls the function aitken to solve this example:

```
x=1:.2:2;
y=ones(size(x))./x;
interpval=aitken(x,y,1.03);
fprintf('interpolated value= %10.8f\n',interpval);
```

Running this script gives the following output:

```
1         1
1.2000    0.8333    0.9750
1.4000    0.7143    0.9786    0.9720
1.6000    0.6250    0.9813    0.9723    0.9713
1.8000    0.5556    0.9833    0.9726    0.9713    0.9710
2.0000    0.5000    0.9850    0.9729    0.9714    0.9711    0.9710

interpolated value= 0.97095439
```

Notice that the first column in this table contains the tabulated x values, the second column the tabulated y values and the remaining columns give successively higher degree polynomial interpolants generated by Aitken's method. The exact value is $y = 0.970873786$; thus the Aitken interpolated value of $y = 0.97095439$ is correct to four decimal places. Linear interpolation gives 0.9750, a much poorer result with an error of approximately 0.2%.

7.3 INTERPOLATION USING SPLINES

The spline is used to connect data points to each other using a curve which appears to the eye to be smooth, either for the purpose of visualisation in design drawings or for interpolation. It has certain advantages over the use of a high-order polynomial which has a tendency to oscillate between data values.

We will begin with an historical example of ship design. Ships' hulls have always curved in a complex manner in two dimensions. Fig. 7.3.1 shows hull sections for a 74 gun British warship, circa 1813. The data points are taken from the original plans and splines have been used to join the data points together smoothly. Each line shows a section of the ship; the innermost line is close to the stern and the outermost is near amidships. The graph gives a clear impression of the way the ship builder chose to reduce the ship's cross-section towards the stern.

Polynomials of varying degrees are used for splines but here we will only consider the cubic spline. The cubic spline is a series of cubic polynomials joining data points or "knots". Suppose we have n data points joined by $n-1$ polynomials. Each cubic polynomial has four unknown coefficients so that there are $4(n-1)$ coefficients to be determined. Obviously each polynomial must pass through the two data points it joins. This will provide $2(n-1)$ equations that must be satisfied. In order that the polynomials join together smoothly, we require both continuity of slope (y') and curvature (y'')

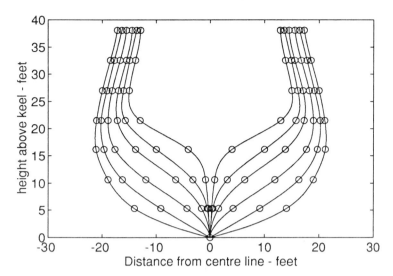

Fig. 7.3.1. Use of splines to define cross-sections of ship's hull.

between adjacent polynomials at the $n-2$ internal data points. This gives $2(n-2)$ extra equations making a total of $4n-6$ equations. With these equations we can determine an identical number of coefficients uniquely and so two further equations are required in order to determine *all* the unknown coefficients. The two remaining conditions can be chosen arbitrarily but usually one of the following is used:

(1) If the slope of the required curve is known at the outer ends, we can impose these two constraints. More often than not, these slopes are not known.

(2) We can make the curvature at the outer ends zero, i.e. $y_1'' = y_n'' = 0$. (These are called natural splines but have no particular advantage.)

(3) We can make the curvature at x_1 and x_n equal to x_2 and x_{n-1} respectively.

(4) We can make the curvature at x_1 a linear extrapolation of the curvature at x_2 and x_3. Similarly we make the curvature at x_n a linear extrapolation of the curvature at x_{n-1} and x_{n-2}.

(5) We can make y''' continuous at x_2 and x_{n-1}. Since at any internal point y, y', y'' and y''' are always made continuous, adding this condition is equivalent to using the same polynomial in the two outer panels. This is called the "not a knot" condition and is used in the MATLAB implementation `spline`.

We will now illustrate the two uses of the MATLAB function `spline` applied to the small set of data given in Table 7.3.1.

Table 7.3.1. Data for spline fit.

x	0	1	2	3	4
y	3	1	0	2	4

Running the script

```
x=[0 1 2 3 4]; y=[3 1 0 2 4]; xval=1.5;
yval=spline(x,y,xval)
```

gives

```
yval =
     0.1719
```

This is perfectly straightforward. There may be some occasions when the user wishes to know the values of the coefficient of the polynomials. In this case the p-p form is output where the abbreviation p-p means piecewise polynomial. For example:

```
x=[0 1 2 3 4]; y=[3 1 0 2 4];
p=spline(x,y);
```

outputs the vector p. The elements of p are interpreted as follows:

p(1)   is the number of polynomials required, i.e. $n - 1$ where n is the number of data points or knots.

p(2) to p(n+1) are the n values of x at which data points are specified.

p(n+2) is the number of unknown coefficients in each polynomial. For a cubic spline it is 4.

The remaining elements of the vector p contain the $4(n - 1)$ coefficients of the polynomials.

In the above example, $n = 5$ so p(1) is 4, p(2), ..., p(6) give the five values of x, p(7) is 4 and p(8) to p(23) give the coefficients of the polynomials as follows:

$$y = p_8 x^3 + p_{12} x^2 + p_{16} x + p_{20} \qquad 0 \le x \le 1$$

$$y = p_9 (x - 1)^3 + p_{13}(x - 1)^2 + p_{17}(x - 1) + p_{21} \qquad 1 \le x \le 2$$

$$y = p_{10}(x - 2)^3 + p_{14}(x - 2)^2 + p_{18}(x - 2) + p_{22} \qquad 2 \le x \le 3$$

$$y = p_{11}(x - 3)^3 + p_{15}(x - 3)^2 + p_{19}(x - 3) + p_{23} \qquad 3 \le x \le 4$$

It is not necessary for the MATLAB user to know the details of how the p-p values are interpreted. MATLAB provides a function ppval which evaluates a composite polynomial provided its pp value is known. If x and y are vectors of data, then running the following script would output identical values for y1 and y2. This is the value of y interpolated from x_i.

```
y1=spline(x,y,xi)
p=spline(x,y);
y2=ppval(p,xi)
```

The following script gives a plot of the spline fit to the data of Table 7.3.1.

```
x=0:4; y=[3 1 0 2 4]; xx=0:.1:4;
yy=spline(x,y,xx);
plot(x,y,'o',xx,yy);
axis([0 4 -1 4]);
xlabel('x'); ylabel('y');
```

Running this script generates Fig. 7.3.2.

In section 7.2 we have shown how polynomials are used in interpolation. However, their use is not always appropriate. When the data points are widely spaced, and when there are sudden changes in the y values, then polynomials can give very poor results. For example, the nine data points in Fig. 7.3.3 are taken from the function

$$y = 2\{1 + \tanh(2x)\} - x/10$$

This function changes abruptly and if an eighth degree polynomial is fitted to the data it oscillates and the path between data points bears no relationship to the true path. In

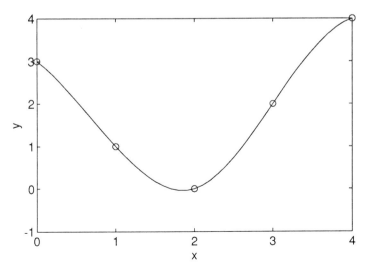

Fig. 7.3.2. Spline fit to the data of Table 7.3.1 denoted by "o".

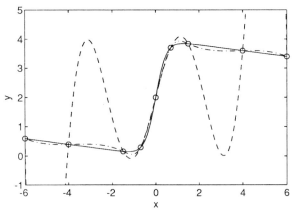

Fig. 7.3.3. The function $y = 2\{1 + \tanh(2x)\} - x/10$ is shown by a
solid line. An eighth degree polynomial fit is shown by a
dashed line and a spline fit is shown by the dot–dashed line.

contrast the spline fit is reasonably smooth and close to the true function.

The reader should note that the MATLAB function `interp1` can also be used to fit splines to data. The call `interp1(x,y,xi,'spline')` is identical to `spline(x,y,xi)`.

7.4 FOURIER ANALYSIS OF DISCRETE DATA

The frequency content of periodic and non-periodic functions can be determined by Fourier analysis. For periodic functions the frequency content is determined from the coefficients of the terms in the well-known Fourier series; for non-periodic functions it is determined from the Fourier transform. In an analogous manner the frequency content of a sequence of data points can be determined by Fourier analysis, in this case from the *discrete* Fourier transform (DFT). The data can come from many sources. For example, the radial forces acting at discrete points around a cylinder constitute a sequence of data which must be periodic. The most frequently occurring form of data is a time series in which the value of some quantity is given at equal intervals of time and for this reason the analysis which follows is developed in terms of the independent variable t which represents time. It must be stressed, however, that the DFT can be applied to any data irrespective of the domain from which it originates. Determining the DFT for a sequence of data points is straightforward, although the computations are tedious. A much more interesting problem arises when the data is sampled from a continuously varying function such as an electrical signal and the DFT is used to estimate the frequency content of the signal. This application makes the DFT an important tool for the scientist or engineer engaged in the interpretation of sampled data, where a knowledge of the frequencies present in the data may give some insight into the mechanism that has generated it.

We begin by defining a periodic function. A function $y(t)$ is periodic if it has the property that for any value of time t, $y(t) = y(t + T)$ where T is the time period, typically measured in seconds. The reciprocal of the period is equal to the frequency, denoted by f and measured in cycles/second. In the SI system of units 1 Hertz (Hz) is defined as 1 cycle/second. If we are concerned with a periodic function $z(x)$, where x is a spatial variable, then for any value of x, $z(x) = z(x + X)$ where X is the spatial period or wavelength, typically measured in metres. The frequency $f = 1/X$ is then measured in cycles/metre.

We now examine how to fit a finite set of trigonometric functions to n data points (t_r, y_r) where $r = 0, 1, 2, ..., n - 1$. We assume the data points are equispaced and the number of data points, n, is even. Data values may be complex but in most practical situations they will be real. The data points are numbered as shown in Fig. 7.4.1. The point following the $(n - 1)$th is assumed to equal the value of the zero point. Thus the DFT assumes the data values are periodic with a period T equal to the range of the data.

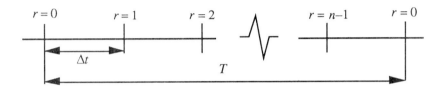

Fig. 7.4.1. Numbering scheme for data points.

Let the relationship between y_r and t_r be given by a finite set of sine and cosine functions thus:

$$y_r = \frac{1}{n}\left(A_0 + \sum_{k=1}^{m-1}\{A_k\cos(2\pi kt_r/T) + B_k\sin(2\pi kt_r/T)\} + A_m\cos(2\pi mt_r/T)\right)$$

(7.4.1)

where $r = 0, 1, 2, ..., n - 1$, $m = n/2$ and T is the range of the data as shown in Fig. 7.4.1. The n coefficients A_0, A_m, A_k and B_k (where $k = 1, 2, ..., m - 1$) must be determined. Since we have n data values and n unknown coefficients, (7.4.1) can be made to fit the data exactly. The factor $1/n$ in (7.4.1) is omitted by some authors and omitting it has the effect of reducing the size of the coefficients A_0, A_m, A_k and B_k by the factor n. The reason for choosing $m + 1$ coefficients multiplied by a cosine function and $m - 1$ coefficients multiplied by a sine function in (7.4.1) will become apparent.

Each sine or cosine term of (7.4.1) represents k complete cycles in the range of the data T. Thus the period of each sine term is T/k, where $k = 1, 2, ..., (m - 1)$, and the period of each cosine term is T/k, where $k = 1, 2, ..., m$. The corresponding frequencies are given by k/T. Thus the frequencies present in (7.4.1) are $1/T, 2/T, ..., m/T$. Letting Δf be the frequency increment and f_{max} be the maximum frequency, then

$$\Delta f = 1/T \tag{7.4.2}$$

and

$$f_{max} = m\Delta f = (n/2)\Delta f = n/(2T) \tag{7.4.3}$$

The data values are equally spaced in the range T and t_r may be expressed as

$$t_r = rT/n, \quad r = 0, 1, 2, ..., n - 1 \tag{7.4.4}$$

Letting Δt be the sampling interval, see Fig. 7.4.1, then

$$\Delta t = T/n \tag{7.4.5}$$

Let T_0 be the period corresponding to f_{max}, the maximum frequency in (7.4.1). Then, from (7.4.3),

$$f_{max} = 1/T_0 = n/(2T)$$

Thus $T = T_0 n/2$. Substituting this relationship in (7.4.5) we have $\Delta t = T_0/2$. This tells us that at the maximum frequency component in the DFT there are two samples of data per cycle. The maximum frequency, f_{max}, is called the Nyquist frequency and the corresponding sampling rate is called the Nyquist sampling rate. A harmonic with a frequency that is exactly equal to the Nyquist frequency cannot be properly detected because at this frequency the DFT has a cosine term but no corresponding sine term. This result has an important implication when data is sampled from a continuously varying function or signal. It implies that there must be *more than two* data samples per cycle at the highest frequency *present in the function or signal*. If there are frequencies in the signal higher than the Nyquist frequency then, because of the periodic nature of the DFT itself, they will appear as frequency components in the DFT at a lower frequency. This phenomenon is called "aliasing". For example, if data is sampled at 0.005 s intervals, i.e. 200 samples/s, then the Nyquist frequency, f_{max}, is 100 Hz. A frequency of 125 Hz in this signal would appear as a frequency component at 75 Hz. A frequency of 225 Hz would appear as 25 Hz. The relationship between frequencies in a signal and the frequency components in the DFT is shown in Fig. 7.4.2. Frequency

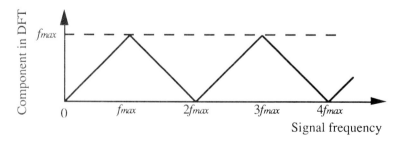

Fig. 7.4.2. Relationship between a signal frequency and its component in the DFT derived by sampling to give a Nyquist frequency f_{max}.

aliasing should be avoided because it makes it difficult or impossible to relate the frequency components in the DFT to their physical causes.

We now return to the task of determining the n coefficients A_0, A_m, A_k and B_k. Replacing $t_r = rT/n$ in (7.4.1) we obtain

$$y_r = \frac{1}{n}\left[A_0 + \sum_{k=1}^{m-1}\{A_k \cos(2\pi kr/n) + B_k \sin(2\pi kr/n)\} + A_m \cos(\pi r)\right] \quad (7.4.6)$$

where $r = 0, 1, 2, ..., n - 1$. It has previously been noted that the coefficients B_0 and B_m are absent from (7.4.1). It is now clear that had we introduced these coefficients they would be multiplied by $\sin(0)$ and $\sin(\pi r)$, both of which are zero.

In (7.4.6) the n unknown coefficients are real. However, (7.4.6) can be expressed more concisely in terms of complex exponentials with complex coefficients. Using the fact that

$$\exp(i2\pi kr/n) = \cos(2\pi kr/n) + i \sin(2\pi kr/n)$$

and

$$\exp\{i2\pi(n - k)r/n\} = \exp(-i2\pi kr/n)$$

where $k = 1, 2, ..., m - 1$, then it can be shown that (7.4.6) reduces to

$$y_r = \frac{1}{n}\sum_{k=0}^{n-1} Y_k \exp(i2\pi kr/n), \quad r = 0, 1, 2, ..., n - 1 \quad (7.4.7)$$

In (7.4.7)

$$Y_0 = A_0 \text{ and } Y_m = A_m, \text{ where } m = n/2$$

$$Y_k = (A_k - iB_k)/2 \text{ and } Y_{n-k} = (A_k + iB_k)/2, \text{ for } k = 1, 2, ..., m - 1$$

Note that if y_r is real then A_k and B_k are also real so that Y_{n-k} is the complex conjugate of Y_k, for $k = 1, 2, ..., (n/2 - 1)$. To find the values of the unknown complex coefficients of (7.4.7) we make use of the following orthogonal property of exponential functions sampled at n equispaced points:

$$\sum_{r=0}^{n-1} \exp(i2\pi rj/n) \exp(-i2\pi rk/n) = \begin{cases} 0 & |j - k| \neq 0, n, 2n, ... \\ n & |j - k| = 0, n, 2n, ... \end{cases} \quad (7.4.8)$$

Multiplying (7.4.7) by $\exp(-i2\pi rj/n)$, summing over the n values of r and then using (7.4.8), an expression for the unknown coefficients can be found thus:

$$Y_k = \sum_{r=0}^{n-1} y_r \exp(-i2\pi kr/n), \quad k = 0, 1, 2, ..., n - 1 \tag{7.4.9}$$

If we let $W_n = \exp(-i2\pi/n)$, where W_n is a complex constant, then (7.4.9) becomes

$$Y_k = \sum_{r=0}^{n-1} y_r W_n^{kr}, \quad k = 0, 1, 2, ..., n - 1 \tag{7.4.10}$$

Alternatively we can write (7.4.10) in matrix notation giving

$$\mathbf{Y} = \mathbf{W}\mathbf{y} \tag{7.4.11}$$

where W_n^{kr} is the element of the $(k + 1)$th row, $(r + 1)$th column of $\mathbf{W}$. Note that $\mathbf{W}$ is an $n \times n$ array of complex coefficients. $\mathbf{Y}$ is a *vector* of the complex Fourier coefficients and in this instance we are departing from our usual convention that emboldened upper case letters represent arrays.

We can obtain the coefficients Y_k from the equispaced data (t_r, y_r) by using (7.4.9), or (7.4.10) or (7.4.11). These equations are alternative statements of the discrete Fourier transform (DFT). Furthermore, by replacing k in (7.4.9) by $k + np$ where p is an integer it can be shown that $Y_{k+np} = Y_k$. Thus the DFT is periodic over the range n. The inverse of the DFT is called the inverse discrete Fourier transform (IDFT) and is implemented by (7.4.7). By replacing r in (7.4.7) by $r + np$ where p is an integer it can be shown that the IDFT is also periodic over the range n. Both y_r and Y_k may be complex although, as has been previously stated, the samples y_r are usually real. These transforms constitute a pair: if the data values are transformed by the DFT to determine the coefficients Y_k then they can be recovered in their entirety by means of the IDFT.

To evaluate the coefficients of the DFT it would appear convenient to use (7.4.11). Although using these equations is satisfactory for small sequences of data, calculating the DFT for n real data points requires $2n^2$ multiplications. Thus, to transform a sequence of 4096 data points would require approximately 33 million multiplications. In 1965 this situation was dramatically changed with the publication of the fast Fourier transform (FFT) algorithm (Cooley and Tukey, 1965). The FFT algorithm is extremely efficient and approximately $2n \log_2 n$ multiplications are required to compute the FFT for real data. With this development, allied to the developments in computing hardware that have occurred in the past 20 years, it is now possible to compute the FFT for a relatively large number of data points on a personal computer.

Many refinements have been made to the basic FFT algorithm since it was first formulated and several variants have been developed. Here one of the simplest forms of the algorithm is outlined.

To develop the basic FFT algorithm one further restriction must be placed on the data. In addition to the data being equispaced, the number of data points must be an integer power of 2. This allows a sequence of data to be successively subdivided. For example, 16 data points can be divided into two sequences of 8, four sequences of 4 and finally eight sequences of only 2 data points. A crucial relationship on which the FFT algorithm is based is now developed from (7.4.9) as follows. Let y_r be the sequence of n data points for which we require the DFT. We can subdivide y_r into two sequences of $n/2$ data points u_r and v_r thus:

$$\left.\begin{array}{l} u_r = y_{2r} \\ \\ v_r = y_{2r+1} \end{array}\right\} r = 0,\ 1,\ 2,\ ...,\ (n/2 - 1) \qquad (7.4.12)$$

Note that alternate points in the original data sequence are placed in different subsets. We now determine the DFTs of the data sets u_r and v_r from (7.4.9), with n replaced by $n/2$, thus:

$$\left.\begin{array}{l} U_k = \displaystyle\sum_{r=0}^{n/2-1} u_r \exp\{-i2\pi kr/(n/2)\} \\ \\ V_k = \displaystyle\sum_{r=0}^{n/2-1} v_r \exp\{-i2\pi kr/(n/2)\} \end{array}\right\} k = 0,\ 1,\ 2,\ ...,\ n/2 - 1$$

The DFT, Y_k, for the original data sequence y_r is given by using (7.4.9) as follows:

$$\begin{aligned} Y_k &= \sum_{r=0}^{n-1} y_r \exp(-i2\pi kr/n) \\ &= \sum_{r=0}^{n/2-1} y_{2r} \exp\{-i2\pi k(2r)/n\} + \sum_{r=0}^{n/2-1} y_{2r+1} \exp\{-i2\pi k(2r+1)/n\} \quad (7.4.13) \end{aligned}$$

where $k = 0,\ 1,\ 2,\ ...,\ n$. Substituting for y_{2r} and y_{2r+1} from (7.4.12) we have

$$Y_k = \sum_{r=0}^{n/2-1} u_r \exp\{-i2\pi kr/(n/2)\} + \exp(-i2\pi k/n) \sum_{r=0}^{n/2-1} v_r \exp\{-i2\pi kr/(n/2)\}$$

Comparing this equation with (7.4.13) we see that

$$Y_k = U_k + \exp(-i2\pi k/n)V_k = U_k + (W_n^k)V_k \qquad (7.4.14)$$

where $W_n^k = \exp(-i2\pi k/n)$ and $k = 0,\ 1,\ 2,\ ...,\ n/2 - 1$.

Equation (7.4.14) provides only half of the required DFT. However, using the fact that U_k and V_k are periodic in k it can be proved that

$$Y_{k+n/2} = U_k - \exp(-i2\pi k/n)V_k = U_k - (W_n^k)V_k \qquad (7.4.15)$$

We can use (7.4.14) and (7.4.15) to determine efficiently the DFT of the original data from the DFTs of subsets composed of alternative points of the original data. Of course, we can determine the DFTs of each subset of data by further subdividing these subsets until the final division leaves subsets consisting of a single data point. For a sequence of data comprising a single data point we see from (7.4.9) with $n = 1$ that the DFT is equal to the value of the single data point. This is essentially how the FFT algorithm works.

In the above discussion we have started from a sequence of data and continuously subdivided it (with alternate points in different subsets) until the subdivisions produce single data points. What we require is a method of starting with single data points and ordering them in such a way that successively combining the DFTs of the subsets ultimately forms the required DFT of the original data. This can be achieved by the "bit reversed algorithm" and we illustrate it and the subsequent stages of the FFT by assuming a sequence of eight data points, y_0 to y_7. To determine the correct order for combining the data we express the subscript denoting the original position of each data point as a binary number and reverse the order of the digits (or bits). This reversed order binary number determines the position of each data point in the reordered sequence and is shown for eight data points in Fig. 7.4.3. The diagram also shows the stages of the FFT algorithm which repeatedly uses (7.4.14) and (7.4.15) thus:

Stage 1: Determine $\mathbf{Y}_{04}$ from Y_0 and Y_4, determine $\mathbf{Y}_{26}$ from Y_2 and Y_6, determine $\mathbf{Y}_{15}$ from Y_1 and Y_5, determine $\mathbf{Y}_{37}$ from Y_3 and Y_7.

Stage 2: Determine $\mathbf{Y}_{0246}$ from $\mathbf{Y}_{04}$ and $\mathbf{Y}_{26}$, determine $\mathbf{Y}_{1357}$ from $\mathbf{Y}_{15}$ and $\mathbf{Y}_{37}$.

Stage 3: Determine $\mathbf{Y}_{01234567}$ from $\mathbf{Y}_{0246}$ and $\mathbf{Y}_{1357}$.

Note that there are three stages in the above process, i.e. $\log_2 n$, where $n = 8$, stages.

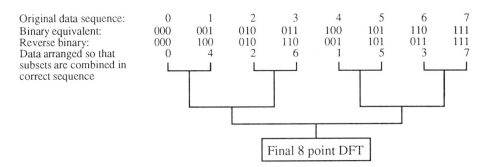

Fig. 7.4.3. Stages in the FFT algorithm.

MATLAB provides both the function fft to determine the DFT of a sequence of data values using the FFT algorithm and the function ifft to determine the IDFT using a slight modification of the FFT algorithm. Thus to determine the DFT of the data in y we use the fft function as the following script illustrates:

```
v=0:15;
y=[2.8 -0.77 -2.2 -3.1 -4.9 -3.2 4.83 -2.5 3.2 ...
 -3.6 -1.1 1.2 -3.2 3.3 -3.4 4.9];
s=sum(y)
Y=fft(y);
[v' Y.']
```

Running this script gives

```
s =
   -7.7400

ans =
         0            -7.7400          [Y(0)]
    1.0000            3.2959 + 8.3851i  [Y(1)]
    2.0000           13.9798 +10.9313i
    3.0000            8.0796 - 6.6525i
    4.0000           -0.2300 + 4.7700i
    5.0000            4.3150 + 6.8308i
    6.0000           14.2202 + 1.4713i
    7.0000          -17.2905 +15.0684i  [Y(n/2 – 1)]
    8.0000           -0.2000            [Y(n/2)]
    9.0000          -17.2905 -15.0684i  [Y(n/2 + 1)]
   10.0000           14.2202 - 1.4713i
   11.0000            4.3150 - 6.8308i
   12.0000           -0.2300 - 4.7700i
   13.0000            8.0796 + 6.6525i
   14.0000           13.9798 -10.9313i
   15.0000            3.2959 - 8.3851i  [Y(n – 1)]
```

We have already noted that for real data, Y_{n-k} is the complex conjugate of Y_k, for $k = 1$, 2, ..., $(n/2 – 1)$. The above results illustrate this relationship and in this case Y_{15}, Y_{14}, ..., Y_9 are the complex conjugates of Y_1, Y_2, ..., Y_7 respectively and provide no extra information. It is also seen that Y_0 is equal to the sum of the original data values y_r.

We now give examples of the use of the fft function to examine the frequency content of data sequences sampled from continuous functions.

Example 1. Determine the DFT of a sequence of 64 equispaced data points that are sampled over a period of 3.2 s from the function $y = 0.5 + 2 \sin(2\pi f_1 t) + \cos(2\pi f_2 t)$, where $f_1 = 3.125$ Hz and $f_2 = 6.25$ Hz. The following script calls the fft function and displays the resulting DFT in various ways:

```
clf; nt=64; T=3.2; dt=T/nt
df=1/T
fmax=(nt/2)*df
t=0:dt:(nt-1)*dt; y=0.5 + 2*sin(2*pi*3.125*t)+cos(2*pi*6.25*t);
f=0:df:(nt-1)*df; Y=fft(y);
figure(1);
subplot(121); bar(real(Y),'r'); axis([0 63 -100 100])
xlabel('index k'); ylabel('real(DFT)')
subplot(122); bar(imag(Y),'r'); axis([0 63 -100 100])
xlabel('index k'); ylabel('imag(DFT)')
fss=0:df:(nt/2-1)*df;
Yss=zeros(1,nt/2); Yss(1:nt/2)=(2/nt)*Y(1:nt/2);
figure(2);
subplot(221); bar(fss,real(Yss),'r'); axis([0 10 -3 3])
xlabel('frequency  Hz'); ylabel('real(DFT)')
subplot(222); bar(fss,imag(Yss),'r'); axis([0 10 -3 3])
xlabel('frequency  Hz'); ylabel('imag(DFT)')
subplot(223); bar(fss,abs(Yss),'r'); axis([0 10 -3 3])
xlabel('frequency  Hz'); ylabel('abs(DFT)')
```

Running the above script gives

```
dt =
    0.0500

df =
    0.3125

fmax =
    10
```

together with Fig. 7.4.4 and Fig. 7.4.5. Note that in the script we have used the bar rather than the plot statement to emphasise the discrete nature of the DFT. Fig. 7.4.4 shows the amplitudes of the 64 real and imaginary components of the DFT plotted against the index number k. Note that components 63 to 33 are the complex conjugates of components 1 to 31. Whilst these plots display the DFT, the amplitude and frequency of the harmonic components in the original signal cannot easily be recognised. To achieve this the DFT must be scaled and displayed as shown in Fig. 7.4.5. In the real part of the DFT there are components at $k = 0$, 20 and 44, each with an amplitude of 32, and in the imaginary part of the DFT there are components at $k = 10$ and 54, with amplitudes of 64 and –64 respectively. For our purpose here we will ignore the components above $k = 32$ (i.e. $k = 44$ and 54) and consider only the components in the range $k = 0, 1, ..., 31$; in this case specifically $k = 0$, 10 and 20. We can convert the DFT index number to frequency by multiplying by Δf (= 0.3125 Hz) to give components at 0 Hz, 3.125 Hz and 6.25 Hz respectively. We now scale the DFT in the range $k = 1, 2, ..., 31$ by dividing it by ($n/2$), in this case by 32. The plots of the 31 scaled DFT components corresponding to frequencies in the range 0 to 9.6875 Hz are shown in Fig. 7.4.5. We now see that the real component at 6.25 Hz has an amplitude of 1 and the imaginary component at 3.125 Hz has an amplitude of 2. These components

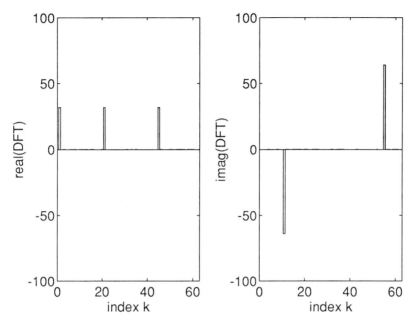

Fig. 7.4.4. Plots of the real and imaginary part of the DFT.

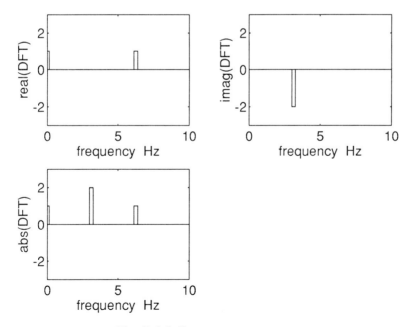

Fig. 7.4.5. Frequency spectra.

correspond to the cosine and sine components in the original signal from which the data was sampled. If we only wish to know the amplitude of the components then we can display the absolute values of the scaled DFT. The component at $f = 0$ Hz is equal to *twice* the mean value of the data, in this case $2 \times 0.5 = 1$. These plots are called frequency spectra or periodograms. If sampling is over an integer number of cycles of the harmonics present in the signal, the amplitude of the component in the scaled DFT will equal the amplitude of these harmonics, as shown in this example. If the sampling is not over an integer number of cycles of any harmonic present in the signal then the component in the DFT closest to the frequency of the harmonic will be reduced in amplitude and spread into other frequencies. This phenomenon is called "smearing" or "leakage" and it is further discussed in problem 7.15.

Example 2. We now determine the spectrum of a sequence of 512 data points sampled over 2.56 seconds from the function

$$y = 0.2 \cos(2\pi f_1 t) + 0.35 \sin(2\pi f_2 t) + 0.3 \sin(2\pi f_3 t) + \text{random noise}$$

where $f_1 = 20$ Hz, $f_2 = 50$ Hz and $f_3 = 70$ Hz. The random noise is normally distributed with a standard deviation of 0.5 and a mean of zero. The following script plots the time series and the DFT scaled by the factor $n/2$.

```
clf
f1=20; f2=50; f3=70; nt=512; T=2; dt=T/nt
df=1/T
fmax=(nt/2)*df; t=0:dt:nt*dt; tt=0:dt/25:nt*dt/50;
y=0.2*cos(2*pi*f1*t)+0.35*sin(2*pi*f2*t)+0.3*sin(2*pi*f3*t);
yy=0.2*cos(2*pi*f1*tt)+0.35*sin(2*pi*f2*tt)+0.3*sin(2*pi*f3*tt);
y=y+0.5*(randn(size(t)));yy=yy+0.5*(randn(size(tt)));
f=0:df:(nt/2-1)*df;
figure(1);
subplot(211), plot(tt,yy)
axis([0 0.04 -3 3])
xlabel('time sec'); ylabel('y')
yf=fft(y); yp=zeros(1,(nt/2));
yp(1:nt/2)=(2/nt)*yf(1:nt/2);
subplot(212), plot(f,abs(yp))
axis([0 fmax 0 0.5])
xlabel('frequency  Hz'); ylabel('abs(DFT)');
```

Running the above script gives

```
dt =
    0.0039

df =
    0.5000
```

together with the following graphical output.

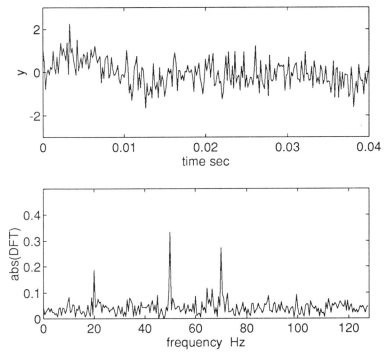

Fig. 7.4.6. Signal and frequency spectrum showing frequency
components at 20, 50 and 70 Hz.

The lower plot of Fig. 7.4.6 shows that random noise in the signal does not prevent the frequency components 20, 50 and 70 Hz from revealing themselves in the spectrum. These components are not obviously visible in the original time series data shown in the upper plot of Fig. 7.4.6.

Example 3. Determine the spectrum of a triangular wave of amplitude ±1 and period 1 second, sampled at 1/32 second intervals over one cycle. The following script outputs the DFT scaled by the factor $n/2$.

```
nt=32;  T=1
dt=T/nt
t=0:dt:T-dt;
df=1/T
fmax=nt/(2*T)
f=0:df:df*(nt/2-1);
y=0.125*[8 7 6 5 4 3 2 1 0 -1 -2 -3 -4 -5 -6 -7 -8 ...
          -7 -6 -5 -4 -3 -2 -1 0 1 2 3 4 5 6 7];
Yss=zeros(1,nt/2);
Y=fft(y);  Yss(1:nt/2)=(2/nt)*Y(1:nt/2);
[f' abs(Yss)']
```

Running this script gives

```
T =
    1

dt =
    0.0312

df =
    1

fmax =

    16

ans =
            0            0
       1.0000       0.8132
       2.0000            0
       3.0000       0.0927
       4.0000            0
       5.0000       0.0352
       6.0000            0
       7.0000       0.0194
       8.0000            0
       9.0000       0.0131
      10.0000            0
      11.0000       0.0100
      12.0000            0
      13.0000       0.0085
      14.0000            0
      15.0000       0.0079
```

The Fourier series for the triangular wave of this example is

$$f(t) = \frac{8}{\pi^2}\left(\cos(2\pi t) + \frac{1}{3^2}\cos(6\pi t) + \frac{1}{5^2}\cos(10\pi t) + \frac{1}{7^2}\cos(14\pi t) + ...\right)$$

The first eight frequency components in the scaled DFT at frequencies 1, 3, 5 Hz etc. are not equal to $8/\pi^2$, $8/(3\pi)^2$, $8/(5\pi)^2$ (i.e. 0.8106, 0.0901, 0.0324) etc., because of the effect of aliasing. A triangular wave contains an infinite number of harmonics and owing to aliasing these appear as components in the DFT as shown in Table 7.4.1. Thus the size of the 3 Hz component in the DFT is $(8/\pi^2)(1/3^2 + 1/29^2 + 1/35^2 + 1/61^2 + ...)$. By summing a large number of terms down the columns of Table 7.4.1 the terms in the DFT are obtained.

Table 7.4.1. Coefficients of aliased harmonics.

f	$3f$	$5f$	$7f$	$9f$	$11f$	$13f$	$15f$
$8/\pi^2$	$8/(3\pi)^2$	$8/(5\pi)^2$	$8/(7\pi)^2$	$8/(9\pi)^2$	$8/(11\pi)^2$	$8/(13\pi)^2$	$8/(15\pi)^2$
$8/(31\pi)^2$	$8/(29\pi)^2$	$8/(27\pi)^2$	$8/(25\pi)^2$	$8/(23\pi)^2$	$8/(21\pi)^2$	$8/(19\pi)^2$	$8/(17\pi)^2$
$8/(33\pi)^2$	$8/(35\pi)^2$	$8/(37\pi)^2$	$8/(39\pi)^2$	$8/(41\pi)^2$	$8/(43\pi)^2$	$8/(45\pi)^2$	$8/(47\pi)^2$
$8/(63\pi)^2$	$8/(61\pi)^2$	$8/(59\pi)^2$	$8/(57\pi)^2$	$8/(55\pi)^2$	$8/(53\pi)^2$	$8/(51\pi)^2$	$8/(49\pi)^2$
$8/(65\pi)^2$	$8/(67\pi)^2$	etc.					

Example 4. Determine the DFT of a sequence of 64 data points sampled from a signal for a period of 8 seconds. The signal has a constant amplitude of 1 unit which, after 0.5 seconds, is switched to zero.

```
clf
y=[ones(1,5) zeros(1,59)];
nt=64; T=8;
dt=T/nt
df=1/T
fmax=(nt/2)*df;
f=0:df:(nt/2-1)*df;
yf=fft(y); yp=zeros(1,nt/2);
yp(1:nt/2)=(2/nt)*yf(1:nt/2);
figure(1); bar(f,abs(yp))
axis([0 fmax 0 0.2])
xlabel('frequency Hz'); ylabel('abs(DFT)')
```

Running this script gives

```
dt =
    0.1250

df =
    0.1250
```

together with the graphical output shown in Fig. 7.4.7. The plot shows that the frequency spectrum is continuous and the largest components are clustered near zero frequency. This is in contrast to the spectrum of Examples 1 and 2 which show sharp peaks due to the presence of periodic components in the original data. Note that because the original signal is a step and not periodic the amplitude of its DFT is dependent on the sampling period.

In this section we have provided examples of how the DFT (computed using the FFT) can be used to study the distribution of frequency components in data. There are

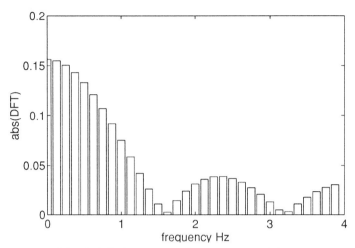

Fig. 7.4.7. Spectrum of a sequence of data.

other applications of the DFT. It can be used for interpolation, as can any procedure that fits mathematical functions to a sequence of data. MATLAB provides the function `interpft` to allow interpolation using the DFT.

The advances in computer hardware have extended the range of problems to which the DFT can usefully be applied and this in turn has encouraged the development of new and powerful variants of the FFT algorithm. A detailed description of the FFT algorithm is given by Brigham (1974) and a straightforward introduction which emphasises the practical problems in using this type of analysis is given by Ramirez (1985).

7.5 FITTING FUNCTIONS TO DATA: LEAST SQUARES CRITERIA

We will now consider the problem of fitting a function to a relatively large amount of data that contains errors. It would not be sensible, nor computationally possible, to fit a very high-order polynomial to a large amount of experimental data as may be done for interpolation. What is required is a function that smooths out fluctuations in the data due to error and reveals any underlying trend. We therefore fit functions that approximate the data according to some criteria. For example, we could adjust the function in order to minimise the maximum error, the sum of the modulus of the errors or the sum of the squares of the errors between the fitted function and the data points. The least squares method is the most widely used of these criteria and we now examine how this process is carried out.

Let the data points be x_i, y_i, where $i = 1, 2, ..., m$. The function to be fitted to the data has the form

$$y = f(x) = a_1\phi_1(x) + a_2\phi_2(x) + ... + a_n\phi_n(x) \tag{7.5.1}$$

where $\phi_i(x)$ are chosen functions and a_i are unknown coefficients. Let the error between this function and the data point (x_k, y_k) be ε_k. Thus

$$\varepsilon_k = y_k - \{a_1\phi_1(x_k) + a_2\phi_2(x_k) + \ldots + a_n\phi_n(x_k)\} \tag{7.5.2}$$

Denoting the sum of the squares of these errors by S we have

$$S = \sum_{i=1}^{m} \left(y_i - \{a_1\phi_1(x_i) + a_2\phi_2(x_i) + \ldots + a_n\phi_n(x_i)\}\right)^2 \tag{7.5.3}$$

We wish to minimise S and to do this we make

$$\frac{\partial S}{\partial a_k} = 0, \ k = 1, 2, \ldots, \ n \tag{7.5.4}$$

Now

$$\frac{\partial S}{\partial a_k} = \sum_{i=1}^{m} 2\left(y_i - \{ a_1\phi_1(x_i) + a_2\phi_2(x_i) + \ldots + a_n\phi_n(x_i)\}\right)\phi_k(x_i) = 0$$

Hence

$$\sum_{i=1}^{m} y_i\phi_k(x_i) - a_1\sum_{i=1}^{m} \phi_1(x_i)\phi_k(x_i) + a_2\sum_{i=1}^{m} \phi_2(x_i)\phi_k(x_i) + \ldots + a_n\sum_{i=1}^{m} \phi_n(x_i)\phi_k(x_i) = 0$$
$$k = 1, 2, \ldots, \ n$$

Rearranging these equations into matrix notation we have

$$\begin{bmatrix} \sum_{i=1}^{m} (\phi_1(x_i))^2 & \sum_{i=1}^{m} \phi_2(x_i)\phi_1(x_i) & \cdots & \sum_{i=1}^{m} \phi_n(x_i)\phi_1(x_i) \\ \sum_{i=1}^{m} \phi_1(x_i)\phi_2(x_i) & \sum_{i=1}^{m} (\phi_2(x_i))^2 & \cdots & \sum_{i=1}^{m} \phi_n(x_i)\phi_2(x_i) \\ \cdots & \cdots & \cdots & \cdots \\ \sum_{i=1}^{m} \phi_1(x_i)\phi_n(x_i) & \sum_{i=1}^{m} \phi_2(x_i)\phi_n(x_i) & \cdots & \sum_{i=1}^{m} (\phi_n(x_i))^2 \end{bmatrix} \begin{bmatrix} a_1 \\ a_2 \\ .. \\ a_n \end{bmatrix} = \begin{bmatrix} \sum_{i=1}^{m} y_i\phi_1(x_i) \\ \sum_{i=1}^{m} y_i\phi_2(x_i) \\ \cdots \\ \sum_{i=1}^{m} y_i\phi_n(x_i) \end{bmatrix}$$

or in matrix form

$$\mathbf{Pa} = \mathbf{b} \tag{7.5.5}$$

where

$$p_{kl} = \sum_{i=1}^{m} \phi_k(x_i)\phi_l(x_i) \text{ and } b_k = \sum_{i=1}^{m} y_i\phi_k(x_i) \tag{7.5.6}$$

7.6 POLYNOMIAL LEAST SQUARES

One of the simplest approaches to curve fitting is based on the polynomial. We may write the general polynomial of degree $n - 1$ in the form

$$y = a_1 + a_2x + a_3x^2 + \dots + a_nx^{n-1} \tag{7.6.1}$$

where $n - 1 \le m$, the number of data points. In practice $n - 1$ is usually much less than m, which means that we are fitting a low degree polynomial through a large amount of data. We must adjust the coefficients a_k to minimise the sum of squares of the errors, as explained in section 7.5. Comparing (7.6.1) with (7.5.1) we see that $\phi_1(x)$ is replaced by 1, $\phi_2(x)$ is replaced by x, $\phi_3(x)$ is replaced by x^2, etc. Thus in general, $\phi_k(x)$ is replaced by x^{k-1}. To determine the coefficients a_k the matrix equation (7.5.5) must be solved. The elements of $\mathbf{P}$ and $\mathbf{b}$ are determined by making the above substitution for $\phi_k(x)$ in (7.5.6) thus:

$$p_{kl} = \sum_{i=1}^{m} x_i^{k+l-2} \text{ and } b_k = \sum_{i=1}^{m} y_i x_i^{k-1} \tag{7.6.2}$$

The function polyfit is provided in the MATLAB toolbox to fit a polynomial of degree n to a sequence of data points using the least squares criteria. The use of this function is now illustrated in the following examples.

Example 1. Fit a cubic polynomial to data generated from $y = 2 + 6x^2 - x^3$ with added random errors. The random errors have a normal distribution with a zero mean value and a standard deviation of 1. The following script calls the MATLAB function polyfit to determine the coefficients of the cubic polynomial followed by polyval to evaluate it for plotting.

```
x=0:.25:6;
y=2+6*x.^2-x.^3;
y=y+randn(size(x));
xx=0:.02:6;
p=polyfit(x,y,3)
yy=polyval(p,xx);
plot(x,y,'o',xx,yy)
axis([0 6 0 40])
xlabel('x'); ylabel('y');
```

Running this script gives the polynomial coefficients in descending powers of x thus:

```
p =
   -0.9678    5.7599    -0.2120    2.5886
```

together with the graph shown in Fig. 7.6.1. The cubic polynomial that fits the data is

$$y = 2.5886 - 0.2120x + 5.7599x^2 - 0.9678x^3$$

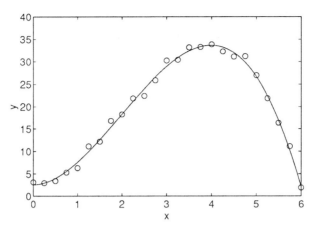

Fig. 7.6.1. Fitting a cubic polynomial to data.

This is different from the original polynomial because of the effect of the random errors in the data. However, the computed polynomial does fit the data as shown in Fig. 7.6.1.

Example 2. Fit a quadratic and cubic polynomial to data generated from $y = e^x$. No random errors are added to the data. The following script solves this problem.

```
x=[0:.25:4]; y=exp(x);
xx=0:.02:4;
p2=polyfit(x,y,2)
yy=polyval(p2,xx);
plot(x,y,'o',xx,yy)
axis([0 4 0 60])
hold on
p3=polyfit(x,y,3)
yy=polyval(p3,xx);
plot(x,y,'o',xx,yy)
xlabel('x'); ylabel('y')
hold off
```

Running this script gives the following coefficients for quadratic and cubic polynomials (in descending powers of x) and plots the data points and the polynomials in Fig. 7.6.2.

```
p2 =
    5.0227    -8.8196     4.3264

p3 =
    1.5648    -4.3659     5.7523     0.2189
```

Although the data points used in the above example are based on an exponential relationship between x and y the cubic polynomial fits the data quite well. This is not always the case as the following example shows.

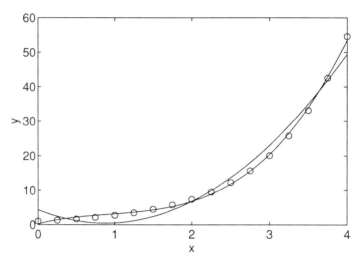

Fig. 7.6.2. Fitting a second and third degree polynomial to data derived from an exponential function. The third-order polynomial fits the data very well.

Example 3. Fit a third- and a fifth-order polynomial to data generated from the function $y = \sin\{1/(x+0.2)\} + 0.2x$ contaminated with random noise, normally distributed with a standard deviation of 0.06 to simulate measurement errors as follows:

```
xs=[0:0.05:0.25  0.25:0.2:4.85];
us=sin(ones(size(xs))./(xs+0.2))+0.2*xs+0.06*randn(size(xs));
save testdata xs us
```

The 30 data values are stored in the file `testdata` so that it can be used in an example in section 7.8. If the data values were regenerated it would be a slightly different sequence because the random noise would change.

The following script loads the data, and fits and plots the least squares polynomial.

```
load testdata
xx=0:.05:5;
p=polyfit(xs,us,3); yy=polyval(p,xx);
plot(xs,us,'o',xx,yy)
hold on
axis([0 5 -2 2])
p=polyfit(xs,us,5); yy=polyval(p,xx);
plot(xx,yy)
xlabel('x'); ylabel('y')
hold off
```

Fig. 7.6.3 shows the result of fitting a third and a fifth degree polynomial to the data and clearly displays the inadequacies of these polynomial approximations. The polynomials oscillate about the points and do not fit the data satisfactorily. In section 7.8 we see that we can improve the fit by using different functions.

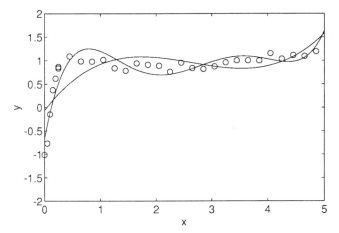

Fig. 7.6.3. Fitting third and fifth degree polynomials to a
sequence of data. Data points are denoted by an "o".

7.7 A PROBLEM WITH POLYNOMIAL LEAST SQUARES FITTING

We have seen in section 7.6 that when we fit a polynomial function to given data the
least squares fit gives rise to equations where the coefficients of the unknowns have the
form given in (7.6.2). That is, the form

$$\sum_{i=1}^{m} x_i^{k+l-2} \tag{7.7.1}$$

We may simplify this to

$$\sum_{i=1}^{m} x_i^{N} \tag{7.7.2}$$

Now systems with coefficients of this form lead to matrices which are extremely ill-
conditioned. This is because for large m the value of (7.7.2) tends to the form of the
coefficients of the Hilbert matrix. A simple justification of this is based on the fact that
the area under the curve $y = x^N$ in the range 0 to 1 can be approximated by dividing the
range 0 to 1 into m intervals and approximating the area in each interval by a rectangle
of width $1/m$ and height x_i^N, where x_i is a point in the interval. For a large number of
intervals this approximation will be very accurate. Hence the area would be given
approximately by

$$\frac{1}{m} \sum_{i=1}^{m} x_i^{N}$$

However, we can find the area simply by integrating x^N between 0 and 1 and the value of this is $1/(N + 1)$. Hence

$$\sum_{i=1}^{m} x_i^N \approx \frac{m}{(N + 1)}$$

Since these are in the form of the coefficients of the Hilbert matrix except for the common factor m we have demonstrated the relationship that the coefficients of the least squares equations have with those of the Hilbert matrix.

In Chapter 2 it is shown that the Hilbert matrix is very ill-conditioned. To illustrate the influence of this on the accuracy of computations we note that the number of decimal places lost when working with an ill-conditioned matrix $\mathbf{A}$ is given approximately by the MATLAB expression `log10(cond(A))`. Thus if we were fitting a fifth-order polynomial the number of decimal places that could be lost may be estimated by `log10(cond(hilb(5)))`. This equals 5.6782, i.e. five or six of the 16 significant digits that MATLAB uses are lost. One way to avoid this problem is to formulate the problem so that no system of linear equations has to be solved. An ingenious way of doing this is to use orthogonal polynomials. We will not describe this method here but refer the reader to Lindfield and Penny (1989).

7.8 GENERAL LEAST SQUARES

We will now provide a MATLAB function which will allow us to fit any given set of functions to data, i.e. we will implement (7.5.5) and (7.5.6). The function `fgenfit` takes the form

```
function c=fgenfit(func,x,y)
if any(size(x)~=size(y))
  disp('X and Y vectors must be the same size')
end
n=length(y);
[p,junk]=feval(func,x(1));
A=zeros(p,p); b=zeros(p,1);
for i=1:n
  [junk,f]=feval(func,x(i));
  for j=1:p
    for k=1:p;
      A(j,k)=A(j,k)+f(j)*f(k);
    end
    b(j)=b(j)+y(i)*f(j);
  end
end
c=A\b;
```

We now use this MATLAB function to fit a set of prescribed functions to data. Consider again Example 3 of section 7.6. We will attempt to fit the data to the function $y = A_0 + A_1\sin\{1/(x + 0.2)\} + A_2x$. This function has been chosen because the data values were originally generated from $y = \sin\{1/(x + 0.2)\} + 0.2x$ contaminated with a normally distributed random noise. We define function f701 which defines the functions $z_1 = 1$, $z_2 = \sin\{1/(x + 0.2)\}$ and $z_3 = x$.

```
function [dof,z]=f701(x)
% x must be a row vector
[junk,n]=size(x); dof=3; z=zeros(dof,n);
z(1,:)=ones(1,n); z(2,:)=sin(ones(1,n)./(x+0.2)); z(3,:)=x;
```

The following script calls the function fgenfit.

```
load testdata
xx=0:.05:5;
c=fgenfit('f701',xs,us)
[junk,p]=feval('f701',xx);
yy=c'*p; plot(xs,us,'o',xx,yy)
axis([0 5 -1.5 1.5])
xlabel('x'); ylabel('y');
```

Running this script gives

```
c =
   -0.0049
    1.0092
    0.2019
```

and the graph of Fig. 7.8.1. The function that fits the data in a least squares sense is given by $y = -0.0049 + 1.0092\sin\{1/(x + 0.2)\} + 0.2019x$. This is very close to the original function; the constant -0.0049 is the mean value of the random noise.

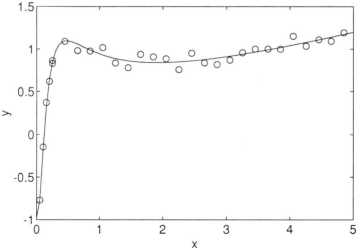

Fig. 7.8.1. Data sampled from the function $y = \sin\{1/(x+ 0.2)\} + 0.2x$.

7.9 TRANSFORMING DATA

A problem can arise when trying to fit functions to data where the relationship between y and the unknown coefficients of the function is non-linear. To illustrate the problem we will begin by simulating experimental data based on the following relationship:

$$y = \frac{1}{\sqrt{(4-x^2)^2 + 0.02}} \qquad (7.9.1)$$

Data values are generated by sampling this function from $x = 1$ to $x = 3$ in increments of 0.05 and small random errors are added to simulate measurement errors. The results of attempting to fit polynomials to these data values are shown in Fig. 7.9.1. The plot shows that as the degree of the polynomial is increased from $n = 4$ to $n = 12$ the polynomial fits the data better in the sense that the total least squars error decreases, but the higher degree polynomials tend to oscillate between the data points. Thus even a 12th degree polynomial does not accurately represent the data, nor does it give us any insight into the underlying mathematical relationship between x and y.

We now consider both the problem and the pitfalls of fitting the function (7.9.2) to the data. In practice this function would be chosen from a knowledge of the suspected relationship between y and x, based on experience or intuition.

$$y = \frac{1}{\sqrt{(a_1 - a_2 x^2)^2 + a_3}} \qquad (7.9.2)$$

The coefficients a_1, a_2 and a_3 must be adjusted to minimise the error between the function and the data. If we compare (7.9.2) with (7.5.1) it is apparent that whilst in (7.5.1) y is a linear combination of the coefficients a_i this is not so in (7.9.2). The fact

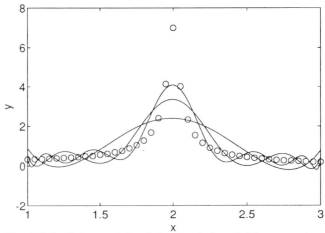

Fig. 7.9.1. Polynomials of degree 4, 8 and 12 attempting to fit a sequence of data indicated by "o" in the graph.

that in (7.5.1) there is a linear relationship between y and the coefficients a_i, makes it possible to develop the system of linear equations (7.5.5) which can be relatively easily generated and solved. When the relation between y and the unknown coefficients is non-linear we can expand the function in a Taylor series and solve the resulting set of non-linear algebraic equations iteratively. This can be slow and unless the starting values used in the iteration are reasonably good, the iteration may not even converge. A further difficulty is that the user must not only supply the function y but also its partial derivatives $\partial y/\partial a_0$, $\partial y/\partial a_1$, etc. Carelessness by the user in evaluating these partial derivatives would have disastrous effects!

An alternative approach to the problem of fitting functions where the relationship between y and the unknown coefficients is non-linear is now considered. This is to transform both the data and the function so that the function has a linear relationship between y and the unknown coefficients. The only difficulty with this is that no general rule can be given to provide a suitable transform; indeed such a transform may not exist. Consider (7.9.2). Letting $Y = 1/y^2$ and $X = x^2$ we have

$$Y = 1/y^2 = (a_1 - a_2x^2)^2 + a_3 = a_2{}^2x^4 - 2a_1a_2x^2 + (a_1{}^2 + a_3)$$

$$= a_2{}^2X^2 - 2a_1a_2X + (a_1{}^2 + a_3)$$

$$= b_1X^2 + b_2X + b_3 \qquad\qquad (7.9.3)$$

Thus Y is a quadratic in X. If the data values are transformed by letting $Y_i = 1/y_i{}^2$ and $X_i = x_i{}^2$ then the process of fitting $Y = f(X)$ to these transformed data values will be a standard least squares polynomial fit giving b_1, b_2 and b_3. Hence the values of a_1, a_2 and a_3 can easily be determined.

We illustrate the above process by considering a sequence of data values related by (7.9.2) to which we have added normally distributed random errors having a zero mean value and a standard deviation of 1%. We can transform these data points using (7.9.3). We must bear in mind that it is the errors in $1/y^2$ rather than y which are being minimised in the least squares process. This may be a problem because it can lead to a great sensitivity to small changes or errors in the data and if the reader experiments with this problem this phenomenon will be demonstrated.

The following script generates the required data, transforms the data points and fits a polynomial to them.

```
clf
x=[1:.05:3];xx=[1:.005:3];
e=0.01*randn(size(x));
y=(ones(size(x))./sqrt((4-x.^2).^2+.02)).*(1+e);
p4=polyfit(x,y,4);yy4=polyval(p4,xx);
p8=polyfit(x,y,8);yy8=polyval(p8,xx);
p12=polyfit(x,y,12);yy12=polyval(p12,xx);
plot(x,y,'o',xx,yy4,xx,yy8,xx,yy12)
axis([1 3 -2 8]); xlabel('x'); ylabel('y')
```

[Script continues...

```
Y=ones(size(x))./y.^2; X=x.^2; XX=xx.^2;
p=polyfit(X,Y,2), YY=polyval(p,XX);
figure(2);
plot(X,Y,'o',XX,YY)
axis([1 9 0 25]); xlabel('X'); ylabel('Y');
yy=ones(size(YY))./sqrt(YY);
figure(3);
plot(x,y,'o',xx,yy)
axis([1 3 -2 8]); xlabel('x'); ylabel('y')
```

Running this script gives the following results:

```
p =
    0.9973    -7.9748    15.9597
```

together with the plot shown in Fig. 7.9.2. From the output of the script we see that the

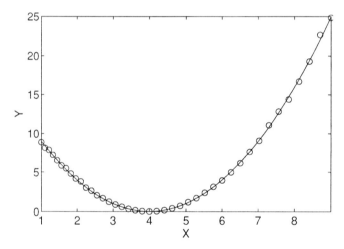

Fig. 7.9.2. Fitting transformed data to a quadratic function.

relationship between X and Y is

$$Y = 0.9973X^2 - 7.9748X + 15.9597$$

We can deduce the values of the unknown coefficients by comparing the above equation with (7.9.3) to give

$a_2^2 = 0.9973$, hence $a_2 = \pm0.9987$. (Take positive value as in (7.9.1).)
$-2a_1a_2 = -7.9748$, hence $a_1 = 3.9928$.
$a_1^2 + a_3 = 15.9597$, hence $a_3 = 0.0173$.

We use these values in the original function (7.9.2) and fit it to the given data. This is shown in Fig. 7.9.3. This function provides a much better fit than the polynomials,

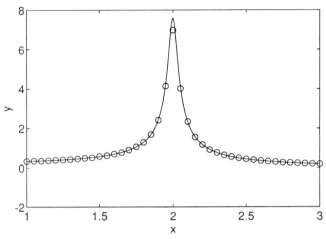

Fig. 7.9.3. Fitting (7.9.2) to the given data.

shown in Fig. 7.9.1. However, even this fit does not pass through the peak value. This is caused by the sensitivity of the process to small random errors in the data. If the random errors are removed the fit is exact. If the script is rerun with the random errors the fit may be worse and if the size of the random errors is increased the process may fail. This is because in the region of $x = 2$ the value of y is essentially only dependent on a_3. This has a small value which may vary in sign.

We now list some functions with a non-linear relationship between y and the coefficients of the function and the corresponding transformations that linearise these relationships.

Relationship	Transform	Linearising substitution
$y = cx^b$	$\log y = \log c + b \log x$	$Y = \log y$ and $X = \log x$
$y = a/(bx+c)$	$1/y = (bx + c)/a$	$Y = 1/y$ and $X = x$
$y = 1/(ax + b)^2$	$1/\sqrt{y} = ax + b$	$Y = 1/\sqrt{y}$ and $X = x$

7.10 SUMMARY

Methods have been described for fitting functions to data for the purposes of interpolation. These have included Aitken's method and spline fits. For periodic data we have examined the fast Fourier transform. Finally, we have discussed least squares approximations to experimental data using polynomial and more general functions.

PROBLEMS

7.1. The following tabulation gives values of the complete elliptic integral

$$E(\alpha) = \int_0^{\pi/2} \sqrt{(1 - \sin^2\alpha \, \sin^2\theta)} \, d\theta$$

$\alpha°$	$E(\alpha)$
0	1.57079
5	1.56780
10	1.55888
15	1.54415
20	1.52379
25	1.49811
30	1.46746

Determine $E(\alpha)$ for $\alpha = 2°$, $13°$ and $27°$ using the MATLAB function aitken.

7.2. Generate a table of values of $f(x) = x^{1.4} - \sqrt{x} + 1/x - 100$ for $x = 20:2:30$. Find the value of x corresponding to $f(x) = 0$ using the MATLAB function aitken. This is an example of inverse interpolation since we are finding the value of x corresponding to a given value of $f(x)$. In particular this gives an approximation to the root of the equation $f(x) = 0$. Compare your solution with that of problem 3.2 in Chapter 3.

7.3. Given $x = -1:0.2:1$, calculate values of y from $y = \sin^2(\pi x/2)$. Using this data:

(i) Generate quadratic and quartic polynomials to fit this data using the least squares MATLAB function polyfit. Display the data and the curve fitted. *Hint*: Example 2 in section 7.6 gives some guidance.

(ii) Fit a cubic spline to the data using the MATLAB function spline. Display the data and the fitted spline. Compare the quality of this spline fit with the two graphs from (i).

7.4. For the data of problem 7.3, determine the values of y for $x = 0.85$ using the MATLAB function interp1 for a linear, spline and cubic interpolating function. Also use the MATLAB function aitken.

7.5. Fit a cubic spline and a fifth degree polynomial to the following data.

x	y
−2	4
0	0
2	−4
3	−30
4	−40
5	−50

Plot the data points, the spline and the polynomial on the same graph. Which curve appears to give the more realistic representation of any underlying function from which the data might have been taken?

7.6. For the data given by the vectors $x = 0:0.25:3$ and

$$y = [6.3806 \ 7.1338 \ 9.1662 \ 11.5545 \ 15.6414 \ 22.7371 \ 32.0696$$
$$47.0756 \ 73.1596 \ 111.4684 \ 175.9895 \ 278.5550 \ 446.4441]$$

fit the following functions:

(i) $f(x) = a + be^x + ce^{2x}$ using the MATLAB function `fgenfit`.

(ii) $f(x) = a + b/(1 + x) + c/(1 + x)^2$ using the MATLAB function `fgenfit`.

(iii) $f(x) = a + bx + cx^2 + dx^3$ using the MATLAB function `polyfit`.

You should plot the three trial functions and the data. How well do these functions fit the data? The data values were in fact generated from $f(x) = 3 + 2e^x + e^{2x}$ with a small amount of random noise added.

7.7. The following values of x and corresponding values of y_u and y_l define an airfoil section:

$$x = [0 \ 0.005 \ 0.0075 \ 0.0125 \ 0.025 \ 0.05 \ 0.1 \ 0.2 \ 0.3 \ 0.4 \ 0.5 \ 0.6 \ 0.7$$
$$0.8 \ 0.9 \ 1]$$

$$y_u = [0 \ 0.0102 \ 0.0134 \ 0.0170 \ 0.0250 \ 0.0376 \ 0.0563 \ 0.0812 \ 0.0962$$
$$0.1035 \ 0.1033 \ 0.0950 \ 0.0802 \ 0.0597 \ 0.0340 \ 0]$$

$$y_l = [0 \ -0.0052 \ -0.0064 \ -0.0063 \ -0.0064 \ -0.0060 \ -0.0045 \ -0.0016$$
$$0.0010 \ 0.0036 \ 0.0070 \ 0.0121 \ 0.0170 \ 0.0199 \ 0.0178 \ 0]$$

The (x, y_u) coordinates define the upper surface and (x, y_l) coordinates define the lower surface. Use the MATLAB function `spline` to fit separate splines to the upper and lower surfaces and plot the results as a single figure.

7.8. Consider the approximation

$$\prod_{p<P}\left(1+\frac{1}{p}\right) \approx C_1 + C_2 \log_e P$$

where the product is taken of all the prime numbers p less than a prime number P. Write a script to generate these products from the list of prime numbers provided and fit the function $C_1 + C_2 \log_e P$ to the points given by the primes P and the corresponding values of the products using the MATLAB function fgenfit.

primes = [2 3 5 7 11 13 17 19 23 29 31 37 41 43 47 53 59 61 67 71 73
79 83 89 97 101 103]

7.9. The gamma function may be approximated by a fifth degree polynomial as

$$\Gamma(x + 1) = a_0 + a_1 x + a_2 x^2 + a_3 x^3 + a_4 x^4 + a_5 x^5$$

Use the MATLAB function gamma to generate values of $\Gamma(x + 1)$ for $x = 0:0.1:1$. Then using the MATLAB function polyfit, fit a fifth degree polynomial to this data. Compare your answers with the approximation for the gamma function given by Abramowitz and Stegun (1965), which gives: $a_0 = 1$, $a_1 = -0.5748666$, $a_2 = 0.9512363$, $a_3 = -0.6998588$, $a_4 = 0.4245549$, $a_5 = -0.1010678$. These coefficients give an accuracy for the gamma function in the range $0 \le x \le 1$ of less than or equal to 5×10^{-5}.

7.10. Generate a table of values of z from the function

$$z(x, y) = 0.5(x^4 - 16x^2 + 5x) + 0.5(y^4 - 16y^2 + 5y)$$

in the range $x = -4:0.2:4$ and $y = -4:0.2:4$. Use this data and the MATLAB function interp2 to interpolate a value for z at $x = y = -2.9035$. Use both linear and cubic interpolation and check your answer by direct substitution in the function. This point gives the global minimum of this function.

7.11. The difference between the mean Sun and the real Sun is called the equation of time. Thus the value of the equation of time $E =$ (mean Sun time – real Sun time). The following table gives E in minutes at 20 equispaced intervals during the year, beginning 1st January.

E = [-3.5 -10.5 -14 -14.25 -9 -4 1 3.5 3 -0.25 -3.5 -6.25 -5.5
-1.75 4 10.5 15 16.25 12.75 6.5]

Use the function interpft to interpolate 300 points and plot E over a period of one year. Then use the command [x,y]=ginput(4) to read from the graph the values of the two minimum and two maximum values of E. At what times do these maxima and minima occur?

7.12. Determine the real and imaginary parts of the DFT, using the MATLAB function fft, for the following periodic data where the 32 data points are sampled at intervals of 0.1 second. Examine the amplitude and frequency of its components. What conclusions can you draw from these results?

$$y = [2 \ -0.404 \ 0.2346 \ 2.6687 \ -1.4142 \ -1.0973 \ 0.8478 \ -2.37 \ 0$$
$$2.37 \ -0.8478 \ 1.0973 \ 1.4142 \ -2.6687 \ -0.2346 \ 0.404 \ -2$$
$$1.8182 \ 1.7654 \ -1.2545 \ 1.4142 \ -0.3169 \ -2.8478 \ 0.9558$$
$$0 \ -0.9558 \ 2.8478 \ 0.3169 \ -1.4142 \ 1.2545 \ -1.7654 \ -1.8182]$$

7.13. Determine the DFT of $y = 32 \sin^5(2\pi ft)$ where $f = 30$ Hz. Use 512 points sampled over 1 second. From the imaginary part of the DFT estimate the coefficients a_0, a_1, a_2 in the relationship

$$32 \sin^5(2\pi ft) = a_0 \sin[2\pi ft] + a_1 \sin[2\pi(3f)t] + a_2 \sin[2\pi(5f)t]$$

Repeat the process for $y = 32 \sin^6(2\pi ft)$ where $f = 30$ Hz. Use 512 points sampled over 1 second. From the real part of the DFT estimate the coefficients b_0, b_1, b_2, b_3 in the relationship

$$32 \sin^6(2\pi ft) = b_0 + b_1 \cos[2\pi(2f)t] + b_2 \cos[2\pi(4f)t] + b_3 \cos[2\pi(6f)t]$$

7.14. Determine the DFT of a set of 512 data points sampled over a 1 second period from

$$y = \sin(2\pi f_1 t) + 2 \sin(2\pi f_2 t)$$

where $f_1 = 30$ Hz and $f_2 = 400$ Hz. Explain why there is a large component in the spectrum at 112 Hz.

7.15. Determine the DFT of a set of 256 data points sampled over 1 second from $y(t) = \sin(2\pi ft)$ for $f = 25$, 30.27 and 35.49 Hz. Plot the absolute value of the DFT against frequency for all three values of f in the same figure. It will be noted that even though the amplitude of the sine function from which the samples are taken is the same in each case the frequency components corresponding to f have different amplitudes. This is because, in the case of the 30.27 and 35.49 Hz wave, the sampling is not over an integer number of periods of y. This phenomenon is known as "leakage" or "smearing" and part of the pure sine wave seems to have smeared into adjacent frequencies. Its effect may be reduced by applying a "window" to the data. The Hanning window is $w(t) = 0.5\{1 - \cos(2\pi t/T)\}$ where T is the sampling period. Multiply $y(t)$ by $w(t)$ and determine the DFT of the resulting data. Plot the absolute value of this DFT against frequency for all three values of f in the same figure. It will be seen that the amplitude variation of the frequency components corresponding to f and the smearing into other frequencies has been reduced significantly.

7.16. The following 32 data points are sampled over a period of 0.0625 second.

$y = [0$ 0.9094 0.4251 −0.6030 −0.6567 0.2247 0.6840 0.1217 −0.5462
        −0.3626 0.3120 0.4655 −0.0575 −0.4373 −0.1537 0.3137
        0.2822 −0.1446 −0.3164 −0.0204 0.2694 0.1439 −0.1702
        −0.2065 0.0536 0.2071 0.0496 −0.1594 −0.1182 0.0853
        0.1441 −0.0078]

(a) Determine the DFT and estimate the frequency of the most significant component present in the data. What is the frequency increment in the DFT?

(b) To the end of the existing data, add an additional 480 zero values thus increasing the number of data points to 512. This process is called "zero padding" and is used to improve the frequency resolution in the DFT. Determine the DFT of the new data set and estimate the frequency of the most significant component. What is the frequency increment in the DFT?

8

Optimisation methods

8.1 INTRODUCTION

The purpose of this chapter is to bring together a selection of algorithms of significant difficulty which have applications in science and engineering. We consider:

(1) the use of interior point methods for solving linear programming problems,

(2) the use of the conjugate gradient method for solving non-linear optimisation problems and systems of linear equations,

(3) the solution of optimisation problems using the genetic algorithm.

It is not our intention to describe fully the theoretical basis for these methods but to give some indication of the ideas that lie behind them. We begin with a discussion of the linear programming problem.

8.2 LINEAR PROGRAMMING PROBLEMS

Linear programming is normally considered to be an operational research (OR) method but has a very wide range of applications. A detailed description of the problem and associated theory is beyond the scope of this text but this information can be obtained from Dantzig (1963). The problem may be expressed in standard form as

Minimise $c^T x$

subject to $Ax = b$ (8.2.1)

and $x \geq 0$

where x is the column vector of n components that we wish to determine. The given constants of the system are provided by an m component column vector b, an $m \times n$ matrix A and an n component column vector c. Clearly all the equations and the function we wish to minimise are linear in form. The problem is an optimisation problem and in general it represents the requirement to minimise a linear function, $c^T x$, called the objective function, subject to satisfying a system of linear equalities.

The importance of this type of problem lies in the fact that it corresponds to the general aim of optimising the use of scarce resources to meet a specific objective. Although we have given the standard form, many other forms of this problem arise which are easily converted to this standard form. For example, the constraints may initially be inequalities and these can be converted to equalities by adding or subtracting additional variables introduced to the problem. The objective may be to maximise the function rather than minimise it. Again this is easily converted by changing the signs of the c coefficients.

Some practical examples where linear programming has been applied are

  (1) the hospital diet problem, requiring food costs to be minimised whilst dietary
      constraints are satisfied,

  (2) the problem of minimising cutting pattern loss,

  (3) the problem of optimising profit subject to constraints on the availability of
      specified materials,

  (4) the problem of optimising the routing of telephone calls.

An important numerical algorithm for solving this problem called the simplex method was introduced by Dantzig (1963). This was applied to wartime problems of troop and material distribution. However, here we consider more recent developments which have provided new algorithms which are theoretically better. These are based on the work of Karmarkar (1984) who produced an algorithm which differed greatly in principle from that of Dantzig. Whilst the theoretical complexity of Dantzig's method is exponential in the number of variables of the problem, some versions of Karmarkar's algorithm have a complexity which is of the order of the cube of the number of variables. It has been reported that for some problems this leads to substantial saving in computational effort. Here we describe an algorithm due to Barnes (1986) that provides an elegant modification of Karmarkar's algorithm but preserves its fundamental principles.

We will not describe the theoretical details of these complex algorithms but it is useful to compare, in broad terms, the nature of the Karmarkar and Dantzig algorithms. The simplex method of Dantzig is best illustrated by considering a simple linear programming problem as follows. In a factory producing electronic components let x_1 be the number of batches of resistors and x_2 the number of batches of capacitors produced. Each batch of resistors manufactured gains 7 units of profit and each batch of capacitors gains 13 units of profit. Each is manufactured in a two-stage process. Stage 1 is limited to 18 units of time per week and stage 2 is limited to 54 units of time per week. A batch of resistors requires 1 unit of time in stage 1 and 5 units of time in stage 2. A batch of capacitors requires 3 units of time in the first stage and 6 in the second. The aim of the manufacturer is to maximise profitability while meeting the time constraints and this leads to the following linear programming problem.

Maximise $z = 7x_1 + 13x_2$ where z is the profit

subject to $x_1 + 3x_2 \leq 18$ stage 1 process
 $5x_1 + 6x_2 \leq 54$ stage 2 process

and $x_1, x_2 \geq 0$

To see how the simplex algorithm works we give a geometric interpretation of this problem in Fig. 8.2.1. In this figure the region lying under the shaded lines and confined by the x_1- and x_2-axes represents the feasible region. This is the region in which all possible solutions to the problem lie. Clearly there are an infinity of such points. Fortunately it can be shown that the only true candidates for the optimum solution are the points that lie at the vertices of the feasible region. In fact we can find this optimum using simple geometric principles. The objective function is represented in Fig. 8.2.1

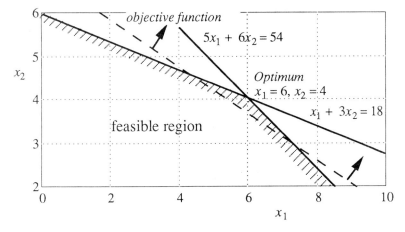

Fig. 8.2.1. The dashed line represents the objective function
and the solid lines the constraints.

by the dashed line of constant slope and variable intercept proportional to the value of the objective function. If we move this line parallel to itself until it just leaves the feasible region it will leave at the vertex which gives the maximum value of the objective function. Clearly beyond this point the values of x_1 and x_2 will no longer satisfy the constraints. For this problem the optimum solution is given by $x_1 = 6, x_2 = 4$ so that the profit $z = 94$ units.

Although this provides a solution for this simple two-variable problem, linear programming problems often involve thousands or hundreds of thousands of variables. For practical problems a well-specified numerical algorithm is required. This is provided by Dantzig's simplex algorithm. We will not describe this in detail here but the general principles of its operation are to generate a sequence of points which correspond mathematically to the vertices of the multidimensional feasible region. The algorithm proceeds from one vertex to another, each time improving the value of the objective function, until the optimum is found. These points are all on the surface of the feasible region and for larger problems there will be a huge number of them.

The algorithm proposed by Karmarkar deals with the linear programming problem in a different way. The algorithm was developed at AT & T to solve very large linear programming problems concerned with routing telephone calls in the Pacific Basin. This algorithm transforms the problem to a more convenient form and then searches through the interior of the feasible region using a good direction of search towards its surface. Because this type of algorithm uses interior points it is often described as an *interior point* method. Since its discovery many improvements and modifications have been made to this algorithm and here we describe a form which, although conceptually complex, leads to a remarkably simple and elegant linear programming algorithm. This formulation was given by Barnes (1986).

Barnes' algorithm may be applied to any linear programming problem once it is converted to the form of (8.2.1). However, one important initial modification is required to ensure the algorithm starts at an interior point $\mathbf{x}^0 > \mathbf{0}$. This modification is achieved by introducing an additional column, i.e. a new last column, to the $\mathbf{A}$ matrix, the elements of which are the $\mathbf{b}$ vector minus the sum of the columns of the $\mathbf{A}$ matrix. We associate an additional variable with this additional column and in order that we do not have a superfluous variable in the solution we introduce an extra element in the vector $\mathbf{c}$. We make the value of this element very large to ensure that the new variable is driven to zero when the optimum is reached. Now we find that $\mathbf{x}^0 = [1\ 1\ 1\ ...\ 1]^T$ will satisfy this set of constraints and clearly $\mathbf{x}^0 > \mathbf{0}$. We now describe the Barnes algorithm:

Step 0:   Assuming n variables in the original problem, set $a(i, n+1) = b(i) - \sum_j a(i, j)$
           and $c(n + 1) = 10000$; $\mathbf{x}^0 = [1\ 1\ 1\ ...\ 1]$; $k = 0$.

Step 1:   Set $\mathbf{D}^k = \text{diag}(\mathbf{x}^k)$ and compute an improved point using the equation

$$\mathbf{x}^{k+1} = \mathbf{x}^k - s(\mathbf{D}^k)^2(\mathbf{c} - \mathbf{A}^T\lambda^k)/\text{norm}((\mathbf{D}^k)(\mathbf{c} - \mathbf{A}^T\lambda^k))$$

where

$$\lambda^k = (\mathbf{A}(\mathbf{D}^k)^2\mathbf{A}^T)^{-1}\,\mathbf{A}(\mathbf{D}^k)^2\,\mathbf{c}$$

The step s is chosen such that

$$s = \min\{\text{norm}((\mathbf{D}^k)(\mathbf{c} - \mathbf{A}^T\lambda^k))/[x_j^k(c_j - \mathbf{A}_j^T\lambda^k)]\} - \alpha$$

where $\mathbf{A}_j$ is the jth column of the matrix $\mathbf{A}$ and α is a small preset constant
value. Here the minimum is taken for the values $(c_j - \mathbf{A}_j^T\lambda^k) > 0$ only. Note
also that λ^k provides an approximation for the solution of the dual problem
(*see below*).

Step 2:   Stop if the *primal* and *dual* values of the objective functions are approximately
           equal. Else set $k = k + 1$ and repeat from step 1.

Note that in step 2 we use an important result in linear programming. This is that every
primal problem (i.e. the original problem) has a corresponding dual problem and if a
solution exists the optimal values of their objective functions are equal. There are
several other termination criteria that could be used and Barnes suggested a more
complex but more reliable one.

The algorithm provides an iterative improvement starting from the initial point $\mathbf{x}^0$
by taking the maximum step which ensures that $\mathbf{x}^k > \mathbf{0}$ in the normalised direction
$(\mathbf{D}^k)^2(\mathbf{c} - \mathbf{A}^T\lambda^k)$. It is this direction which is the crucial element of the algorithm. This
direction is a projection of the objective function coefficients into the constraint space.
For a proof that this direction will reduce the objective function, while ensuring the
constraints are satisfied, the reader is referred to Barnes (1986).

The reader should be warned that this algorithm is deceptively simple. In fact the
computation of the direction is very difficult for large problems. This is because the
algorithm requires the solution of an extremely ill-conditioned equation system. Many
alternatives have been suggested for finding the direction of search including the use
of a conjugate gradient method which is discussed in section 8.4 of this chapter. The
MATLAB function barnes provided here solves the ill-conditioned equation system in
a direct manner using the MATLAB \ operator. The function barnes is easily modified
to use the conjugate gradient solver given in section 8.4.

```
function [xsol,basic]=barnes(A,b,c,tol)
x2=[ ]; x=[ ]; [m n]=size(A);
%Set up initial problem
aplus1=b-sum(A(1:m,:)')'; cplus1=1000000;
A=[A aplus1]; c=[c cplus1];
B=[ ]; n=n+1;
x0=ones(1,n)'; x=x0;
alpha = .0001; lambda=zeros(1,m)'; iter=0;
%Main step
while abs(c*x-lambda'*b)>tol
  x2=x.*x; D=diag(x); D2=diag(x2);
  AD2=A*D2;
  lambda=(AD2*A')\(AD2*c');
  dualres=c'-A'*lambda;
  normres=norm(D*dualres);
  for i=1:n
    if dualres(i)>0
      ratio(i)=normres/(x(i)*(c(i)-A(:,i)'*lambda));
    else
      ratio(i)=inf;
    end
  end
  R=min(ratio)-alpha;
  x1=x-R*D2*dualres/normres;
  x=x1; basiscount=0; B=[ ]; basic=[ ]; cb=[ ];
  for k=1:n
    if x(k)>tol
      basiscount=basiscount+1;
      basic=[basic k];
    end
  end
  %Only used if problem non-degenerate
  if basiscount==m
    for k=basic
      B=[B A(:,k)]; cb=[cb c(k)];
    end
    primalsol=b'/B'; xsol=primalsol;
    break
  end
  iter=iter+1;
end;
objective=c*x
```

We now solve the linear programming problem

$$\text{Maximise} \quad z = 2x_1 + x_2 + 4x_3$$

$$\text{subject to} \quad \begin{aligned} x_1 + x_2 + x_3 &\leq 7 \\ x_1 + 2x_2 + 3x_3 &\leq 12 \end{aligned}$$

and the non-negativity constraints $x_1, x_2, x_3 \geq 0$

This linear programming problem can be easily transformed to the standard form by adding new positive-valued variables, called slack variables, to the left-hand sides of the inequalities and changing the signs of the coefficients in the objective function so that it is converted to a minimisation problem subject to equality constraints thus:

$$\text{Minimise} \quad -z = -2x_1 - x_2 - 4x_3$$

$$\text{subject to} \quad \begin{aligned} x_1 + x_2 + x_3 + x_4 &= 7 \\ x_1 + 2x_2 + 3x_3 + x_5 &= 12 \end{aligned}$$

and the non-negativity constraints $x_1, x_2, x_3, x_4, x_5 \geq 0$

The variables x_4 and x_5 are called the slack variables and they represent the difference between the available resources and the resources used. Note that if the constraints had been of the form greater than or equal to zero then we would subtract slack variables to produce equality. Thus we have

$$\mathbf{A} = \begin{bmatrix} 1 & 1 & 1 & 1 & 0 \\ 1 & 2 & 3 & 0 & 1 \end{bmatrix} \quad \mathbf{b} = \begin{bmatrix} 7 \\ 12 \end{bmatrix}$$

$$\mathbf{c} = [-2 \ -1 \ -4 \ 0 \ 0]$$

We use the following script to solve this problem.

```
c=[-2 -1 -4 0 0];
a=[1 1 1 1 0;1 2 3 0 1 ]; b=[7 12]';
[xsol,ind]=barnes(a,b,c,.00005);
i=1;fprintf('\nSolution is:');
for j=ind
  fprintf('\nx(%1.0f)=%8.4f\',j,xsol(i));
  i=i+1;
end;
fprintf('\nOther variables are zero\n')
```

Running this script provides the result

```
objective =
  -19.0000

Solution is:
x(1)=  4.5000
x(3)=  2.5000
Other variables are zero
```

This solution illustrates an important theorem of linear programming. The number of non-zero primal variables is at most equal to the number of independent constraints (excluding non-negativity constraints). In this problem there are only two main constraints. Thus there are only two non-zero variables, x_1 and x_3. The slack variables

x_4 and x_5 are zero and so is x_2.

Having examined the process for solving linear optimisation problems we now consider a method which is widely used to solve non-linear optimisation problems.

8.3 THE CONJUGATE GRADIENT METHOD

We will confine ourselves to solving the problem

Minimise $f(\mathbf{x})$ for all $\mathbf{x} \in R^n$

where $f(\mathbf{x})$ is a non-linear function of $\mathbf{x}$ and $\mathbf{x}$ is an n component column vector. This is called a non-linear unconstrained optimisation problem. These problems arise in many applications, for example in neural network problems where an important aim is to find weights in a network which minimise the difference between the input of the network and the required output.

The standard approach for solving this problem is to assume an initial approximation $\mathbf{x}^0$ and then to proceed to an improved approximation by using an iterative formula of the form

$$\mathbf{x}^{k+1} = \mathbf{x}^k + s\mathbf{d}^k \quad \text{for } k = 0, 1, 2, \dots \tag{8.3.1}$$

Clearly to use this formula we must determine values for the scalar s and the vector $\mathbf{d}^k$. The vector $\mathbf{d}^k$ represents a direction of search and the scalar s determines how far we should step in this direction. A vast literature has grown up which has examined the problem of choosing the best direction and the best step size to solve this problem efficiently. For example, see Adby and Dempster (1974). A simple choice for a direction of search is to take $\mathbf{d}^k$ as the negative gradient vector at the point $\mathbf{x}^k$. For a sufficiently small step value this can be shown to guarantee a reduction in the function value. This leads to an algorithm of the form

$$\mathbf{x}^{k+1} = \mathbf{x}^k - s\nabla f(\mathbf{x}^k) \quad \text{for } k = 0, 1, 2, \dots \tag{8.3.2}$$

where $\nabla f(\mathbf{x}) = (\partial f/\partial x_1, \partial f/\partial x_2, \dots, \partial f/\partial x_n)$ and s is a small constant value. This is called the steepest descent algorithm. The minimum is reached when the gradient is zero, as in the ordinary calculus approach. We will also assume that there exists only one local minimum which we wish to find in the range considered. The problem with this method is that although it will reduce the function value, the step may be very small and therefore the algorithm is very slow. An alternative approach is to choose the step which gives the maximum reduction in the function value in the current direction. This may be described formally as

For each k find the value of s that minimises $f(\mathbf{x}^k - s\nabla f(\mathbf{x}^k))$ \hspace{1em} (8.3.3)

This procedure is known as a line-search. The reader will note that this is also a minimisation problem. However, since $\mathbf{x}^k$ will be known it is a minimisation problem in one variable only, the step size s. Although it is a difficult problem, numerical procedures are available to solve it. Equations (8.3.2) and (8.3.3) provide a workable algorithm but it is still slow. One reason for this poor performance lies in our choice of direction $\nabla f(\mathbf{x}^k)$.

Consider the function we wish to minimise in (8.3.3). Clearly the value of s which will minimise $f(\mathbf{x}^k - s\nabla f(\mathbf{x}^k))$ will be such that the derivative of $f(\mathbf{x}^k - s\nabla f(\mathbf{x}^k))$ with respect to s will be zero. Now differentiating $f(\mathbf{x}^k - s\nabla f(\mathbf{x}^k))$ with respect to s gives

$$\frac{\mathrm{d}f\left(\mathbf{x}^k - s\nabla f\left(\mathbf{x}^k\right)\right)}{\mathrm{d}s} = -\left(\nabla f\left(\mathbf{x}^k\right)\right)^{\mathrm{T}} \nabla f\left(\mathbf{x}^k\right) = 0 \tag{8.3.4}$$

This shows that the successive directions of search are orthogonal. This is not the best way of getting from our original approximation to the optimum value since the changes in direction are so large.

The conjugate gradient method takes a combination of the previous direction and the new direction to approach the optimum more directly. It uses the same step size choice procedure given by (8.3.3) so we must now consider how the direction vector is chosen in the conjugate gradient method. Let $\mathbf{g}^{k+1} = \nabla f(\mathbf{x}^{k+1})$ so that the basic formula for the conjugate gradient direction is

$$\mathbf{d}^{k+1} = -\mathbf{g}^{k+1} + \beta \mathbf{d}^k \tag{8.3.5}$$

Thus the current direction of search is a combination of the current negative gradient plus a scalar β times the previous direction of search. The crucial question is: how is the value of β to be determined? The criterion used is that successive directions of search should be conjugate. This means that $(\mathbf{d}^{k+1})^{\mathrm{T}} \mathbf{A} \, \mathbf{d}^k = 0$ for some specified matrix $\mathbf{A}$.

This apparently obscure choice of requirement can be shown to lead to desirable convergence properties for the conjugate gradient method. In particular it has the property that the optimum of a positive definite quadratic function of n variables can be found in n or less steps. In the case of a quadratic, $\mathbf{A}$ is the matrix of coefficients of the squared and cross product terms. It can be shown that the requirement of conjugacy leads to a value for β given by (8.3.6):

$$\beta = (\mathbf{g}^{k+1})^{\mathrm{T}} \, \mathbf{g}^{k+1}/(\mathbf{g}^k)^{\mathrm{T}} \, \mathbf{g}^k \tag{8.3.6}$$

Now (8.3.1), (8.3.3), (8.3.5) and (8.3.6) lead to the conjugate gradient algorithm given by Fletcher and Reeves (1964) which has the form

Step 0: Input value for $\mathbf{x}^0$ and accuracy ε. Set $k = 0$ and compute $\mathbf{d}^k = -\nabla f(\mathbf{x}^k)$.

Step 1: Determine s_k which is the value of s that minimises $f(\mathbf{x}^k + s\mathbf{d}^k)$.

         Calculate $\mathbf{x}^{k+1}$ where $\mathbf{x}^{k+1} = \mathbf{x}^k + s_k\mathbf{d}^k$ and compute $\mathbf{g}^{k+1} = \nabla f(\mathbf{x}^{k+1})$.

         If norm$(\mathbf{g}^{k+1}) < \varepsilon$ then terminate with solution $\mathbf{x}^{k+1}$ else go to step 2.

Step 2: Calculate new conjugate direction $\mathbf{d}^{k+1}$ where

$$\mathbf{d}^{k+1} = -\mathbf{g}^{k+1} + \beta\mathbf{d}^k \text{ and } \beta = (\mathbf{g}^{k+1})^\mathrm{T}\,\mathbf{g}^{k+1}/\{(\mathbf{g}^k)^\mathrm{T}\,\mathbf{g}^k\}$$

Step 3: $k = k + 1$; go to step 1.

Note that in other forms of this algorithm the steps 1, 2 and 3 are repeated n times and then restarted with a steepest descent step from step 0. A MATLAB function for this method is given below.

```
function res=mincg(f,derf,ftau,x,tol)
global p1 d1
n=size(x); noiter=0;
%Calculate initial gradient
df= feval(derf,x);
%main loop
while norm(df)>tol
  noiter=noiter+1;
  df= feval(derf,x); d1=df;
  %Inner loop
  for inner=1:n
    p1=x;
    % Linear search accuracy = 0.001; reduce for greater accuracy.
    tau=fmin(ftau,-2,2,[0 .001]);
    % calculate new x
    x1=x-tau*d1;
    %Save previous gradient
    dfp=df;
    %Calculate new gradient
    df= feval(derf,x1);
    %Update x and d
    d=d1; x=x1;
    %Conjugate gradient method
    beta=(df'*df)/(dfp'*dfp);
    d1=-df+beta*d;
  end;
end;
res=x1;
disp('Solution'); disp(x1);
disp('Gradient'); disp(df);
```

Notice that the toolbox MATLAB function fmin is used in the function mincg to perform the single variable minimisation to find the best step value. It is important to note that the function mincg requires three input functions which must be supplied by the user. They are the function to be minimised, the partial derivatives of this function and the line-search function. An example of the use of mincg is given below.

The function to be optimised, which is taken from Styblinski and Tang (1990), is

$$f(x_1, x_2) = (x_1^4 - 16x_1^2 + 5x_1)/2 + (x_2^4 - 16x_2^2 + 5x_2)/2$$

The MATLAB function f801 which defines it is as follows:

```
function fv=f801(x);
fv=0.5*(x(1)^4-16*x(1)^2+5*x(1)) +...
  0.5*(x(2)^4-16*x(2)^2+5*x(2));
```

A MATLAB function for the partial derivatives is f801pd:

```
function pd=f801pd(x);
pd=zeros(size(x));
pd(1)=4*x(1)^3-32*x(1)+5;
pd(2)=4*x(2)^3-32*x(2)+5;
```

The MATLAB line-search function ftau2cg is defined as

```
function ftauv=ftau2cg(tau);
global p1 d1
q1=p1-tau*d1;
ftauv=feval('f801',q1);
```

To test the mincg function we use the following simple MATLAB script:

```
global p1 d1
x0=[0 1]';
x1=mincg('f801','f801pd','ftau2cg',x0,0.000005);
```

The results of running this script are

```
Solution
   -2.9035
   -2.9035

Gradient
   1.0e-05 *
   0.1914
   0.0806
```

It is interesting to see the function we have optimised and we give both a three-dimensional and contour plot in Fig. 8.3.1 and Fig. 8.3.2. The latter includes a plot of

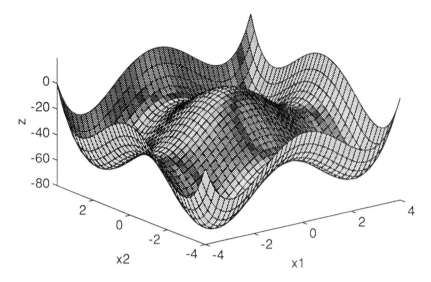

Fig. 8.3.1. Three-dimensional plot of the function f801.

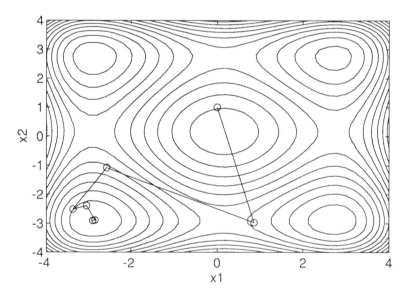

Fig. 8.3.2. Contour plot of the function f801 showing the location of four local minima. The conjugate gradient algorithm has found the one in the lower left-hand corner. The search path taken by the algorithm is also shown.

the iterates and shows the path taken to reach the optimum solution. The script used to obtain these graphs is

```
clf
[x,y]=meshgrid(-4.0:0.2:4.0,-4.0:0.2:4.0);
z=0.5*(x.^4-16*x.^2+5*x)+0.5*(y.^4-16*y.^2+5*y);
figure(1)
surfl(x,y,z);
axis([-4 4 -4 4 -80 20])
xlabel('x1'); ylabel('x2'); zlabel('z');
x1=[0 0.8618 -2.5776 -3.3682 -3.0484 -2.8512 -2.9056 -2.9035];
y1=[1 -2.9644 -1.0750 -2.5092 -2.3899 -2.8824 -2.8968 -2.9035];
figure(2)
contour(z,15,-4.0:0.2:4.0,-4.0:0.2:4.0);
xlabel('x1'); ylabel('x2');
hold on
plot(x1,y1,x1,y1,'o')
xlabel('x1');ylabel('x2');
hold off
```

In this script the vectors x1 and y1 contain the iterates for the conjugate gradient solution of the given function. These values were obtained by running a modified version of the mincg function separately. The minimum we have obtained is in fact the smallest of the four local minima that exist for this function. However, this result was fortuitous; all that the conjugate gradient method is able to do is to find one of the four local minima and even this is not guaranteed for all problems. The conjugate gradient method, because of its limited storage requirements, is one of the key algorithms used in neural network problems as part of the back propagation algorithm but it has many other applications.

It should be noted that a MATLAB optimisation toolbox is available and this provides a range of optimisation procedures.

We now consider another application of the conjugate gradient algorithm, this time for the solution of linear systems of equations which have a special form.

8.4 THE CONJUGATE GRADIENT ALGORITHM FOR SOLVING LINEAR EQUATION SYSTEMS

We now apply the conjugate gradient algorithm to minimise a positive definite quadratic function which has the standard form

$$f(\mathbf{x}) = (\mathbf{x}^T\mathbf{A}\mathbf{x})/2 + \mathbf{p}^T\mathbf{x} + q \qquad (8.4.1)$$

Here $\mathbf{x}$ and $\mathbf{p}$ are n component column vectors, $\mathbf{A}$ is an $n \times n$ positive definite symmetric matrix and q is a scalar. The minimum value of $f(\mathbf{x})$ will be such that the gradient of $f(\mathbf{x})$ is zero. However, the gradient is easily found by direct differentiation as

$$\nabla f(\mathbf{x}) = \mathbf{A}\mathbf{x} + \mathbf{p} = \mathbf{0} \qquad (8.4.2)$$

Thus finding the minimum is equivalent to solving this system of linear equations which becomes, on letting $\mathbf{b} = -\mathbf{p}$,

$$\mathbf{Ax} = \mathbf{b} \tag{8.4.3}$$

Since we can use the conjugate gradient method to find the minimum of (8.4.1) then we can use it to solve the equivalent system of linear equations (8.4.3). The conjugate gradient method provides a powerful method for solving linear equation systems with positive definite symmetric matrices and follows quite closely the algorithm we have described for solving non-linear optimisation problems. However, the line-search is greatly simplified and the value of gradient can be computed within the algorithm in this case. The algorithm takes the form

Step 0: $k = 0$: $\mathbf{x}^k = \mathbf{0}$, $\mathbf{g}^k = \mathbf{b}$, $\mu^k = \mathbf{b}^T\mathbf{b}$, $\mathbf{d}^k = -\mathbf{g}^k$,

Step 1: while system is not satisfied

$$\mathbf{q}^k = \mathbf{A}\mathbf{d}^k,\ \ r^k = (\mathbf{d}^k)^T\mathbf{q}^k,\ \ s^k = \mu^k/r^k$$

$$\mathbf{x}^{k+1} = \mathbf{x}^k + s^k\mathbf{d}^k,\ \ \mathbf{g}^{k+1} = \mathbf{g}^k + s^k\mathbf{q}^k$$

$$t^k = (\mathbf{g}^{k+1})^T\,\mathbf{q}^k,\ \ \beta^k = t^k/r^k$$
$$\mathbf{d}^{k+1} = -\mathbf{g}^{k+1} + \beta^k\,\mathbf{d}^k,\ \ \mu^{k+1} = \beta^k\,\mu^k$$
$$k = k + 1$$
end

Notice that the values of the gradient $\mathbf{g}$ and the step s are calculated directly and no MATLAB function or user-defined function is required.

The MATLAB function solvercg implements this algorithm and utilises the stopping procedure suggested by Karmarkar and Ramakrishnan (1991). See this paper and also Golub and Van Loan (1989).

```
function xdash=solvercg(a,b,n,tol)
xdash=[ ]; gdash=[ ]; ddash=[ ]; qdash=[ ]; q=[ ];
mxitr=n*n;
xdash=zeros(n,1); gdash=-b; ddash=-gdash;
muinit=b'*b;
stop_criterion1=1;
k=0;
mu=muinit;
```

[*Script continues...*

```
% main stage
while (stop_criterion1==1)
  qdash=a*ddash;
  q=qdash;
  r=ddash'*q;
  if (r==0),
    error('r=0,divide by 0!!!')
  end
  s=mu/r;
  xdash=xdash+s*ddash; gdash=gdash+s*q;
  t=gdash'*qdash; beta=t/r;
  ddash=-gdash+beta*ddash;
  mu=beta*mu;
  k=k+1;
  val=a*xdash;
  if((1-val'*b/(norm(val)*norm(b)))<=tol)&(mu/muinit<=tol),
    stop_criterion1=0;
  end
  if(k>mxitr)
    stop_criterion1=0;
  end
end
```

The script below generates a system of 10 equations with randomly selected elements on which this algorithm can be tested:

```
n=10; tol=1e-8;
A=10*rand(n); b=10*rand(n,1);
ada=A*A';
% To ensure a symmetric positive definite matrix.
sol=solvercg(ada,b,n,tol);
disp('Solution of system is:');
disp(sol);
accuracy=norm(ada*sol-b);
fprintf('Norm of residuals =%12.9f\n',accuracy);
```

Running this script gives the following results:

```
Solution of system is:
    0.0949
    0.5715
   -0.5377
    0.3948
   -0.2400
    0.2679
    0.4775
    0.0177
   -0.6292
   -0.4733

Norm of residuals = 0.000000003
```

We note that the norm of the residuals is very small. For ill-conditioned matrices it is necessary to use some kind of preconditioner which reduces the condition number of the matrix, otherwise the method becomes too slow. Karmarkar and Ramakrishnan (1991) have used a preconditioned conjugate gradient method as part of an interior point algorithm to solve linear programming problems with 5000 rows and 333000 columns.

8.5 GENETIC ALGORITHMS

In this section we introduce the ideas on which genetic algorithms are based and provide a group of MATLAB functions which implement the key features of a genetic algorithm. These are applied to the solution of some optimisation problems. It is beyond the scope of this book to give a detailed account of this rapidly developing field of study and the reader is referred to the excellent text of Goldberg (1989).

Genetic algorithms have been the subject of considerable interest in recent years since they appear to provide a robust search procedure for solving difficult problems. The striking feature of these algorithms is that they are based on ideas from the science of genetics and the process of natural selection. This cross fertilisation from one field of science to another has led to stimulating and fruitful applications in many fields and particularly in computer science.

We will now describe the genetic algorithm in the terminology used in the field and then explain how this relates to an optimisation problem. The genetic algorithm works with an initial *population* which may, for example, correspond to numerical values of a particular variable. The size of this population may vary and is generally related to the problem under consideration. The members of this population are usually strings of zeros and ones, i.e. binary strings. For example, a small initial or first-generation population may take the form

    1000010
    1110000
    1010101
    1111001
    1000001

In practice the population may be far larger than this and the strings longer. The strings themselves may be the encoded values of a variable or variables that we are examining. This initial population is generated randomly and we can use the terminology of genetics to characterise it. Each string in the population corresponds to a *chromosome* and each binary element of the string to a *gene*. A new population must now develop from this initial population and to do this we implement the analogue of specific fundamental genetic processes. These are:

(1) Reproduction based on fitness
(2) Crossover
(3) Mutation

A set of chromosomes is selected at the reproduction stage based on natural selection. Thus members of the population are chosen for reproduction on the basis of their fitness defined according to some specified criteria. The fittest are given a greater probability of reproducing in proportion to the value of their fitness.

The actual process of *mating* is implemented by using the simple idea of crossover. This means that two members of the population exchange genes. There are many ways of implementing this crossover, for example having a single crossover point or many crossover points. These crossover points are selected randomly. A simple crossover is illustrated below for two chromosomes selected according to fitness. Here we have randomly selected a crossover point after the fourth digit.

    1110 |000
    1010 |101

After crossover this gives

    1110 |101
    1010 |000

Applying this procedure to our original population we produce a new generation. The final process is mutation. Here we randomly change a particular gene in a particular chromosome. Thus a 0 may be changed to a 1 or vice versa. The process of mutation in a genetic algorithm occurs very rarely and hence this probability of a change in a string is kept very low.

Having described the basic principles of a genetic algorithm we will now illustrate how it may be applied by considering a simple optimisation problem and in so doing fill in some of the details to show how a genetic algorithm may be implemented. A manufacturer wishes to produce a wall mounting container which consists of a hemisphere surmounted by a cylinder of fixed height. The height of the cylinder is fixed but the common radius of the cylinder and hemisphere may be varied between 2 and 4. The manufacturer wishes to find the radius value which maximises the volume of the container. This is a simple problem and the optimum radius is 4. However, it will serve to illustrate how the genetic algorithm may be applied.

We can formulate this as an optimisation problem by taking r as the common radius of the cylinder and hemisphere and h as the height of the cylinder. Taking $h = 2$ units leads to the formula

$$\text{Maximise } v = 2\pi r^3/3 + 2\pi r^2 \tag{8.5.1}$$

where $2 \leq r \leq 4$.

The first problem we must consider is how to transform this problem so that the genetic algorithm can be applied directly. First we must generate an initial set of strings to constitute the initial population. The number of bits in each string, i.e. the string length, will limit the accuracy with which we can find the solution to the problem so it must be chosen with care. In addition we must select the size of the initial population; again this must be chosen with care since a large initial population will increase the time taken to implement the steps of the algorithm. A large population may be unnecessary since the algorithm automatically generates new members of the population in the process of searching the region. The MATLAB function genbin is used to generate such an initial population and takes the form

```
function chromosome=genbin(bitl,numchrom)
% Generates binary pop'n size numchrom with bit length bitl
maxchros=2^bitl;
if numchrom>=maxchros
  numchrom=maxchros;
end
for k=1:numchrom
  for bd=1:bitl
    if rand>=.5
      chromosome(k,bd)=1;
    else
      chromosome(k,bd)=0;
    end;
  end;
end;
```

This function can be defined more succinctly by using the MATLAB round statement thus:

```
function chromosome=genbin(bitl,numchrom)
maxchros=2^bitl;
if numchrom>=maxchros
  numchrom=maxchros;
end
chromosome=round(rand(numchrom,bitl))
```

To generate an initial population of five chromosomes, each with six genes, we call this function as

```
»chroms=genbin(6,5)

chroms =
     0     1     1     1     0     0
     1     1     1     1     0     1
     1     0     0     1     0     0
     0     0     0     0     1     1
     0     1     1     1     0     1
```

Since we are interested in values of r in the range 2 to 4 we must be able to transform these binary strings to values in the range 2 to 4. This is achieved by using the MATLAB function binvreal which converts a binary value to a real value in the required range.

```
function rval=binvreal(chrom,a,b)
%converts chromosome to real value in range a to b
[pop bitlength]=size(chrom);
maxchrom=2^bitlength-1;
realel=chrom.*((2*ones(1,bitlength)).^fliplr([0:bitlength-1]));
tot=sum(realel);
rval=a+tot*(b-a)/maxchrom;
```

We now call this function to convert the population generated above:

```
»for i=1:5, rval(i)=binvreal(chroms(i,:),2,4); end;
»rval
rval =
    2.8889     3.9365     3.1429     2.0952     2.9206
```

As expected, these values are in the range 2 to 4 and provide the initial population of values for r. However, these values tell us nothing about their fitness and to discover this we must judge them against some fitness criteria. In this case the choice is easy since our objective is to maximise the value of the function (8.5.1). We simply find the values of our objective function (8.5.1) for these values of r. We must define our function as a MATLAB function and it takes the form

```
function fv=f802(x)
fv=pi*x.*x.*(.66667*x+2);
```

Now we use this to evaluate fitness by replacing x by the values rval thus

```
»fit=f802(rval)
fit =
  102.9330   225.1246   127.0806    46.8480   105.7749
```

Notice at this stage that the total fitness is

```
»sum(fit)
ans =
  607.7611
```

So the fittest is the value 3.9365 with a fitness value 225.1246 which corresponds to string or population member 2. Fortuitously this is a very good result.

The next stage is reproduction when the strings are copied according to their fitness. Thus there will be a higher probability of more of the fittest chromosomes in the mating pool. This process of selection is more complex and is based on a process which simulates the use of a roulette wheel. The percentage of the roulette wheel that is allocated to a particular string is directly proportional to the fitness of the string. For the above fitness vector fit the percentages can be calculated from

```
»percent=fit/sum(fit)*100
percent =
   16.9364   37.0416   20.9096   7.7083   17.4040

»sum(percent)
ans =
   100
```

Thus conceptually we spin a roulette wheel on which the strings 1 to 5 have the percentages of area 16.9364, 37.0416, 20.9096, 7.7083 and 17.4040. Thus these chromosomes or strings have this chance of being selected. This is implemented by the function selectga as follows:

```
function newchrom=selectga(criteria,chrom,a,b)
% Selects best chromosomes for next generation using criteria
[pop bitlength]=size(chrom);
fit=[ ];
%calculate fitness
[fit,fitot]=fitness(criteria,chrom,a,b);
for chromnum=1:pop
  sval(chromnum)=sum(fit(1,1:chromnum));
end;
%select according to fitness
parname=[ ];
for i=1:pop
  rval=floor(fitot*rand);
  if rval<sval(1)
    parname=[parname 1];
  else
    for j=1:pop-1
      sl=sval(j); su=sval(j)+fit(j+1);
      if (rval>=sl) & (rval<=su)
        parname=[parname j+1];
      end
    end
  end
end
newchrom(1:pop,:)=chrom(parname,:);
```

We can now use this function to perform the reproduction stage thus:

```
»matepool=selectga('f802',chroms,2,4)
matepool
     1     1     1     1     0     1
     1     1     1     1     0     1
     0     1     1     1     0     0
     0     1     1     1     0     1
     0     1     1     1     0     0
```

The fitness is calculated using the following MATLAB function:

```
function [fit,fitot]=fitness(criteria,chrom,a,b)
% Calculates fitness of set of chromosomes in range a to b.
[pop bitl]=size(chrom);
for k=1:pop
  v(k)=binvreal(chrom(k,:),a,b);
  fit(k)=feval(criteria,v(k));
end;
fitot=sum(fit);
```

We can use this to obtain the fitness of the new population as follows:

```
»fitness('f802',matepool,2,4)
ans =
  225.1246   225.1246   102.9330   105.7749   102.9330

»sum(ans)
ans =
  761.8902
```

Notice the substantial increase in overall fitness.

We can now mate the members of this population but we mate only a proportion of them, in this case 60% or 0.6. The function which carries this out is matesome defined as follows:

```
function chrom1=matesome(chrom,matenum)
mateind=[ ]; chrom1=chrom;
[pop bitlength]=size(chrom); ind=1:pop;
u=floor(pop*matenum);
if floor(u/2)~=u/2,u=u-1;end;
%select percentage to mate randomly
while length(mateind)~=u
  i=round(rand*pop);
  if i==0,i=1;end;
  if ind(i)~=-1
    mateind=[mateind i]; ind(i)=-1;
  end
end
%perform single point crossover
for i=1:2:u-1
  splitpos=floor(rand*bitlength);
  if splitpos==0, splitpos=1; end;
  i1=mateind(i); i2=mateind(i+1);
  tempgene=chrom(i1,splitpos+1:bitlength);
  chrom1(i1,splitpos+1:bitlength)=chrom(i2,splitpos+1:bitlength);
  chrom1(i2,splitpos+1:bitlength)=tempgene;
end;
```

We now use this function to mate the strings in the new population matepool:

```
»newgen=matesome(matepool,.6)
newgen =
     1     1     1     1     0     1
     1     1     1     1     0     0
     0     1     1     1     0     1
     0     1     1     1     0     1
     0     1     1     1     0     0

»fitness('f802',newgen,2,4)
ans =
   225.1246   220.4945   105.7749   105.7749   102.9330

»sum(ans)
ans =
   760.1018
```

Notice that the total fitness has not improved and indeed, at this stage, we cannot expect improvements every time.

Finally we perform a mutation before repeating this same cycle of steps. This is implemented by the function mutate as follows:

```
function chrom=mutate(chrom,mu)
%mutate small portion of chromosomes
[pop bitlength]=size(chrom);
for i=1:pop
  for j=1:bitlength
    if rand<=mu
      if chrom(i,j)==1
        chrom(i,j)=0;
      else
        chrom(i,j)=1;
      end
    end
  end
end
```

This is called with a very small value for mu and a population of this size is unlikely to be changed in just one generation. A call of this function is given below:

```
»mutate(newgen,.005)
ans =
     1     1     1     1     0     1
     1     1     1     1     0     0
     0     1     1     1     0     1
     0     1     1     1     0     1
     0     1     1     1     0     0
```

This completes the production of a new generation. The process is now repeated using the new generation and subsequently repeated for many generations.

The function optga includes all these steps in one function and is defined as follows:

```
function [xval,maxf]=optga(fun,range,bits,pop,gens,mu,matenum)
%GA algorithm
newpop=[]; a=range(1); b=range(2)
newpop=genbin(bits,pop);
for i=1:gens
   selpop=selectga(fun,newpop,a,b);
   newgen=matesome(selpop,matenum);
   newgen1=mutate(newgen,mu);
   newpop=newgen1;
end
[fit,fitot]=fitness(fun,newpop,a,b);
[maxf,mostfit]=max(fit);
xval-binvreal(newpop(mostfit,:),a,b);
```

Now applying this function to solve our original problem we specify the range of x from 2 to 4, use 8 bit length chromosomes and an initial population of 10. The process is continued for 20 generations with a mutation probability of 0.005 and a mating proportion of 0.6. Note that matenum must be greater than zero and less than or equal to one. Thus

```
»[x f]=optga('f802',[2 4],8,10,20,.005,.6)
x =
    3.8667

f =
  215.0202
```

Since the exact solution is $x = 4$ this is a reasonable result. Fig. 8.5.1 gives a graphical representation of the progress of the genetic algorithm. It should be noted that each run

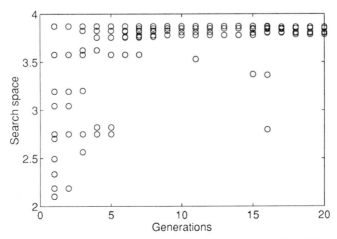

Fig. 8.5.1. Each member of the population is represented by "o". Successive generations of the population concentrate towards the value 4 approximately.

of the genetic algorithm can produce a different result because of the random nature of the process. In addition the number of distinct values in the search space is limited by the chromosome length. In this example the chromosome length is 8 bits giving 2^8 or 256 divisions. Thus the range of r from 2 to 4 is divided into 256 divisions, each equal to 0.0078125.

We will now discuss the philosophy and theory behind this process and the real problems to which genetic algorithms may be applied. The reason why a genetic algorithm differs from a simple direct search procedure is that it involves two special features: crossover and mutation. Thus starting from an initial population the algorithm develops new generations which rapidly explore the region of interest. This is useful for difficult optimisation problems and in particular for those where we wish to find the global maximum or minimum of a function which has many local maxima and minima. In this case standard optimisation methods such as the conjugate gradient method of Fletcher and Reeves can locate only the local optimum. However, a genetic algorithm may locate the global optimum although this is not guaranteed. This is due to the way it explores the region of interest avoiding getting stuck at a particular local minimum. We will not consider the theoretical justification in detail but describe the key result only.

We first introduce the concept of *schemata*. If we study the structure of the strings produced by a genetic algorithm certain patterns of behaviour begin to emerge. Strings which have high fitness values often have common features, a particular combination of binary elements. For example, the fittest strings may have the common feature that they start with 11 and end with 0 or always have the middle three elements 0. We can represent strings with this structure by 11*****0 and ***000** where the asterisks represent "wild card" elements which may be either 0 or 1. These structures are called schemata and essentially they identify the common features of a set of strings. The reason why a particular schema is interesting is because we wish to study the propagation of such strings which have this structure and are associated with high values of fitness. The length of a schema is the distance between the outermost specified gene values. The order of a schemata is the number of positions specified by 0 or 1. For example:

string	order	length
**********1	1	1
******10*1**	3	4
10******	2	2
00******101	5	11
11**00	4	6

It is clear that schemata which are defined by sub-strings of short length are less likely to be affected by crossover and therefore propagate through the generations unchanged.

We can now state the *fundamental theorem of genetic algorithms,* due to Holland, in terms of these schemata. This states that schemata of short length and low order with above average fitness are propagated in exponentially increasing numbers throughout the generations. The ones with below average fitness die away exponentially. This key result explains some of the success of genetic algorithms.

We now provide some further examples which apply the MATLAB genetic algorithm function optga to a specific optimisation routine.

Example 1. Determine the maximum of the following function in the range $x = 0$ to $x = 1$.

$$f(x) = e^x + \sin(3\pi x)$$

The function f803 is defined as

```
function v=f803(x)
v=exp(x)+sin(3*pi*x);
```

Calling optga with this function we have

```
»[x f]=optga('f803',[0 1],8,40,50,.005,.6)
x =
    0.8627

f =
    3.3315
```

We now apply the supplied MATLAB function fmin to solve this problem. Note that for use with fmin the function f803 has been modified by including a minus sign, thereby negating the function since fmin is performing minimisation.

```
»fmin('f803',0,1)
ans =
    0.8602

»f803(.8602)
ans =
   -3.3317
```

This shows that the results are similar although the function fmin is faster than the genetic algorithm.

Example 2. A more demanding problem is to minimise the function

$$f(x) = 10 + [1/\{(x - 0.16)^2 + 0.1\}]\sin(1/x)$$

This is defined by f804 thus:

```
function v=f804(x)
[m,n]=size(x);
p=ones(m,n);
v=10+(p./((x-.16).^2+.1)).*sin(p./x);
```

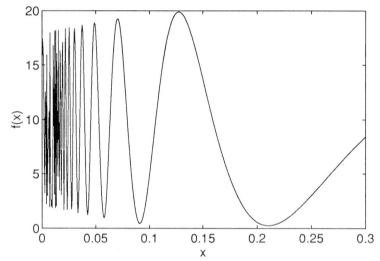

Fig. 8.5.2. Plot of the function $10 + [1/\{(x - 0.16)^2 + 0.1\}]\sin(1/x)$
showing many local optimum values.

Calling `optga` with this function gives the following:

```
»[x, f]=optga('f804',[0.001 0.3],8,10,40,.005,.6)
x =
    0.1206

f =
    18.9180
```

Fig. 8.5.2 illustrates the difficulty of this problem and shows that the result is a reasonable one.

Genetic algorithms are a developing area of research and many amendments could be made to the functions which we have supplied to implement a genetic algorithm. For example, the roulette wheel selection can be implemented in many different ways; crossover can be changed to multi-point crossover or other alternatives. It will often be noticed that a genetic algorithm is slow in execution but it should be remembered that it is best applied to difficult problems, for example those which have multiple optima and where the global optimum is required. Since standard algorithms often fail in these cases the extra time taken by the genetic algorithm is worth while. There are many applications of genetic algorithms which we have not considered and the function `optga` only works for positive-valued functions in one independent variable. It would be easy to extend it to deal with two variable functions and some of these will be given as exercises for the reader.

8.6 SUMMARY

In this chapter we have introduced a number of more advanced areas of numerical analysis which are still topics for active research. MATLAB functions are provided to allow the reader to explore the problems more deeply. The structure of the algorithms is well reflected by the structure of the MATLAB functions.

PROBLEMS

8.1. Use the function $barnes$ to solve the following problem:

$$\text{Minimise} \quad z = 5x_1 + 7x_2 + 10x_3$$

$$\text{subject to} \quad \begin{aligned} x_1 + x_2 + x_3 &\geq 4 \\ x_1 + 2x_2 + 4x_3 &\geq 5 \end{aligned}$$

$$\text{and} \quad x_1, x_2, x_3 \geq 0$$

8.2. Consider the following problem:

$$\text{Maximise} \quad p = 4y_1 + 5y_2$$

$$\text{subject to} \quad \begin{aligned} y_1 + y_2 &\leq 5 \\ y_1 + 2y_2 &\leq 7 \\ y_1 + 4y_2 &\leq 10 \end{aligned}$$

and the non-negativity constraints $y_1, y_2 \geq 0$

By introducing slack variables and subtracting one from each equality write the constraints as equalities. Hence apply the function $barnes$ to solve this problem. Notice that the optimum value of p for this problem is equal to the optimum value of z in problem 8.1. Problem 8.2 is called the dual of problem 8.1. This is an example of an important theorem that the optimum of the objective function of a problem and its dual are equal.

8.3. Maximise $\qquad z = 2u_1 - 4u_2 + 4u_3$

$$\text{subject to} \quad \begin{aligned} u_1 + 2u_2 + u_3 &\leq 30 \\ u_1 + u_2 \phantom{{}+ u_3} &= 10 \\ u_1 + u_2 + u_3 &\geq 8 \end{aligned}$$

and the non-negativity constraints $u_1, u_2, u_3 \geq 0$

Hint: Remember to use slack variables to ensure that the main constraints are equalities.

8.4. Use the function mincg, with tolerance 0.005, to minimise Rosenbrock's function

$$f(x, y) = 100(x^2 - y)^2 + (1 - x)^2$$

starting with the initial approximation $x = 0.5$, $y = 0.5$ and using a line-search accuracy 10 times the machine precision in fmin. To obtain an impression of how this function varies, plot it in the range $0 \leq x \leq 2$, $0 \leq y \leq 2$.

8.5. Use the function mincg, with tolerance 0.00005, to minimise the five variable function

$$z = 0.5(x_1^4 - 16x_1^2 + 5x_1) + 0.5(x_2^4 - 16x_2^2 + 5x_2) + ...$$
$$+ (x_3 - 1)^2 + (x_4 - 1)^2 + (x_5 - 1)^2$$

Use $x_1 = 1$, $x_2 = 2$, $x_3 = 0$, $x_4 = 2$ and $x_5 = 3$ for the starting values in mincg. Experiment further with other starting values.

8.6. Use the function solvercg to solve the matrix equation $\mathbf{Ax} = \mathbf{b}$ where

$$\mathbf{A} = \begin{bmatrix} 5 & 4 & 1 & 1 \\ 4 & 5 & 1 & 1 \\ 1 & 1 & 4 & 2 \\ 1 & 1 & 2 & 4 \end{bmatrix} \quad \mathbf{b} = \begin{bmatrix} 1 \\ 2 \\ 3 \\ 4 \end{bmatrix}$$

Check the accuracy of the solution by finding the value of norm(b-Ax).

8.7. Maximise the function $y = 1/\{(x-1)^2 + 2\}$ in the range $x = 0$ to 2 using the function optga. Use different initial population sizes, mutation rates and numbers of generations. Notice that this is not a simple exercise since for each set of conditions it will be necessary to solve the problem several times to take account of the random nature of the process. Given that the optimum value of the function is 0.5, plot the error in the optimum value of the function for each run under a particular set of parameters. Then change one of the parameters and repeat the process. Differences in the plots may or may not be discernible.

8.8. Plot the function $z = x^2 + y^2$ in the range $0 \leq x \leq 2$ and $0 \leq y \leq 2$. The genetic algorithm given in section 8.5 may be applied to maximise functions in more than one variable. Use the MATLAB function optga to determine the maximum value of the function given above. In order to do this you must modify the fitness function so that the first half of the chromosome corresponds to values of x and the second half to values of y and these chromosomes must map to the values of x and y. For example, if an 8 bit (gene) chromosome 10010111 is split into two parts 1001 and 0111 then it would convert to $x = 9$ and $y = 7$.

Appendix 1 – Matrix algebra

A1.1 INTRODUCTION

For the convenience of the reader we bring together some of the important definitions and relationships of matrix algebra. Since many MATLAB functions and operators act on matrices and arrays it is important that the user of MATLAB should feel at ease with matrix notation and matrix algebra. MATLAB is an ideal environment in which to experiment and learn matrix algebra. Whilst MATLAB cannot provide a formal proof of any relationship it does allow the user to verify results and rapidly gain experience in matrix manipulation. In this appendix only definitions and results are provided. For proofs and further explanation, the reader is recommended to consult Golub and Van Loan (1989).

A1.2 MATRICES AND VECTORS

A matrix is a rectangular array of elements which in itself cannot be evaluated. An element of a matrix can be a real or complex number, an algebraic expression or another matrix. Normally matrices are enclosed in square brackets, round brackets or braces. In this text square brackets are used. A complete matrix is denoted by an emboldened character. For example:

$$
\mathbf{A} = \begin{bmatrix} 3 & -2 \\ -2 & 4 \end{bmatrix}
\quad
\mathbf{B} = \begin{bmatrix} \mathbf{A} & \mathbf{A} & 2\mathbf{A} \\ \mathbf{A} & -\mathbf{A} & \mathbf{A} \end{bmatrix}
\quad \therefore \;
\mathbf{B} = \begin{bmatrix}
3 & -2 & 3 & -2 & 6 & -4 \\
-2 & 4 & -2 & 4 & -4 & 8 \\
3 & -2 & -3 & 2 & 3 & -2 \\
-2 & 4 & 2 & -4 & -2 & 4
\end{bmatrix}
$$

$$\mathbf{x} = \begin{bmatrix} 11 \\ -3 \\ 7 \end{bmatrix} \quad \mathbf{e} = \begin{bmatrix} (2+3i) & (p^2+q) & (-4+7i) & (3-4i) \end{bmatrix}$$

where $i = \sqrt{(-1)}$. In the above examples **A** is a 2 x 2 square matrix with two rows and two columns of real coefficients. It also has the property of being a symmetric matrix, see section A1.7. The matrix **B** is built up from the matrix **A** and so **B** is a 4 x 6 real matrix. The matrix **x** is a 3 x 1 matrix and is usually called a column vector and **e** is a 1 x 4 complex matrix, usually called a row vector. Note that **e** has the algebraic expression $p^2 + q$ for its second element. In this vector each element is enclosed in round brackets to clarify its structure. Enclosing an element in round brackets is not a requirement.

If we wish to refer to a particular element in a matrix we use subscript notation: the first subscript denotes the row, the second the column. In the case of the row and column vectors it is conventional to use a single subscript. Thus, in the examples above,

$$a_{21} = -2, \ b_{25} = -4,$$
$$x_2 = -3, \ e_4 = 3 - 4i$$

Note also that although **A** and **B** are upper case letters it is conventional to refer to their elements by lower case letters. In general the element in the ith row and jth column of **A** is denoted by a_{ij}.

A1.3 SOME SPECIAL MATRICES

The identity matrix. The identity matrix denoted by **I** has unit values along the leading diagonal and zeros elsewhere. The leading diagonal is the diagonal of elements from the top left to the bottom right of the matrix. For example:

$$\mathbf{I}_2 = \begin{bmatrix} 1 & 0 \\ 0 & 1 \end{bmatrix}, \quad \mathbf{I}_3 = \begin{bmatrix} 1 & 0 & 0 \\ 0 & 1 & 0 \\ 0 & 0 & 1 \end{bmatrix}$$

The subscript indicating the size of the matrix is usually omitted. The identity matrix behaves rather like the scalar quantity 1. In particular, pre- or postmultiplying a matrix by **I** does not change it.

Diagonal matrix. Such a matrix is square and composed of zero elements except for some of the elements on the leading diagonal, thus:

$$\mathbf{A} = \begin{bmatrix} 4 & 0 & 0 & 0 \\ 0 & -2 & 0 & 0 \\ 0 & 0 & 0 & 0 \\ 0 & 0 & 0 & 9 \end{bmatrix}, \quad \mathbf{B} = \begin{bmatrix} 12 & 0 & 0 \\ 0 & -2 & 0 \\ 0 & 0 & -6 \end{bmatrix}$$

Tridiagonal matrix. This square matrix has all zero elements except in the leading diagonal and the diagonals immediately above and below it. Thus, using 'x' to denote a non-zero element:

$$\mathbf{A} = \begin{bmatrix} x & x & 0 & 0 & 0 \\ x & x & x & 0 & 0 \\ 0 & x & x & x & 0 \\ 0 & 0 & x & x & x \\ 0 & 0 & 0 & x & x \end{bmatrix}$$

Triangular and Hessenberg matrices. A lower triangular matrix has non-zero elements only on and below the leading diagonal. An upper triangular matrix has non-zero elements on and above the leading diagonal. The Hessenberg matrix is similar to the triangular matrix except that it also has non-zero elements on the diagonals adjacent to the leading diagonal.

$$\begin{bmatrix} x & x & x & x & x \\ 0 & x & x & x & x \\ 0 & 0 & x & x & x \\ 0 & 0 & 0 & x & x \\ 0 & 0 & 0 & 0 & x \end{bmatrix}, \quad \begin{bmatrix} x & 0 & 0 & 0 & 0 \\ x & x & 0 & 0 & 0 \\ x & x & x & 0 & 0 \\ x & x & x & x & 0 \\ x & x & x & x & x \end{bmatrix}, \quad \begin{bmatrix} x & x & x & x & x \\ x & x & x & x & x \\ 0 & x & x & x & x \\ 0 & 0 & x & x & x \\ 0 & 0 & 0 & x & x \end{bmatrix}$$

The first matrix is upper triangular, the second is lower triangular and the last is an upper Hessenberg matrix.

A1.4 DETERMINANTS

The determinant of $\mathbf{A}$ is written $|\mathbf{A}|$ or $\det(\mathbf{A})$. For a 2×2 array we define its determinant as follows:

$$\text{If } \mathbf{A} = \begin{bmatrix} a_{11} & a_{12} \\ a_{21} & a_{22} \end{bmatrix} \text{ then } \det(\mathbf{A}) = \begin{vmatrix} a_{11} & a_{12} \\ a_{21} & a_{22} \end{vmatrix} = a_{11}a_{22} - a_{21}a_{12} \quad \text{(A1.1)}$$

In general for an $n \times n$ array, $\mathbf{A}$, cofactors $C_{ij} = (-1)^{i+j}\Delta_{ij}$ can be defined. In this definition Δ_{ij} is the determinant formed from $\mathbf{A}$ when the elements of the ith row and jth column are deleted. Δ_{ij} is called the minor of $\mathbf{A}$. Then

$$\det(\mathbf{A}) = \sum_{k=1}^{n} a_{ik}C_{ik} \quad \text{(A1.2)}$$

This is known as an expansion along the ith row. Frequently the first row is used. This equation replaces the problem of evaluating one $n \times n$ determinant $\mathbf{A}$ by the evaluation of n, $(n-1) \times (n-1)$ determinants. The process can be continued until the cofactors are reduced to 2×2 determinants. Then the formula (A1.1) is used. This is the formal definition for the determinant of $\mathbf{A}$ but it is not a computationally efficient procedure.

A1.5 MATRIX OPERATIONS

Matrix transposition. In this operation the rows and columns of a matrix are interchanged or transposed. The transposition of a real matrix $\mathbf{A}$ is denoted by $\mathbf{A}^T$. For example:

$$\mathbf{A} = \begin{bmatrix} 1 & -2 & 4 \\ 2 & 1 & 7 \end{bmatrix}, \quad \mathbf{A}^T = \begin{bmatrix} 1 & 2 \\ -2 & 1 \\ 4 & 7 \end{bmatrix} \text{ and } \mathbf{x} = \begin{bmatrix} 1 \\ 2 \\ 3 \end{bmatrix}, \quad \mathbf{x}^T = \begin{bmatrix} 1 & 2 & 3 \end{bmatrix}$$

Note that a square matrix remains square when it is transposed and a column vector transposes into a row vector and vice versa.

Matrix addition and subtraction. This is done by adding or subtracting corresponding elements in the matrices. Thus

$$\begin{bmatrix} 1 & 3 \\ -4 & 5 \end{bmatrix} + \begin{bmatrix} 5 & -4 \\ 6 & 6 \end{bmatrix} = \begin{bmatrix} 6 & -1 \\ 2 & 11 \end{bmatrix}, \quad \begin{bmatrix} -4 \\ 6 \\ 11 \end{bmatrix} - \begin{bmatrix} 3 \\ -3 \\ 2 \end{bmatrix} = \begin{bmatrix} -7 \\ 9 \\ 9 \end{bmatrix}$$

It is apparent that only matrices with the same number of rows and the same number of columns can be added and subtracted. In general, if $\mathbf{A} = \mathbf{B} + \mathbf{C}$ then $a_{ij} = b_{ij} + c_{ij}$.

Scalar multiplication. Every element of a matrix is multiplied by a scalar quantity. Thus if $\mathbf{A} = s\mathbf{B}$, where s is a scalar, then $a_{ij} = sb_{ij}$.

Matrix multiplication. We can only multiply two matrices $\mathbf{B}$ and $\mathbf{C}$ together if the number of columns in $\mathbf{B}$ is equal to the number of rows in $\mathbf{C}$. Such matrices are said to be conformable. If $\mathbf{B}$ is a $p \times q$ matrix and $\mathbf{C}$ is a $q \times r$ matrix then we can determine the product $\mathbf{BC}$ and the result, $\mathbf{A}$, will be a $p \times r$ matrix. Because the order of matrix multiplication is important we say that $\mathbf{B}$ premultiplies $\mathbf{C}$ or $\mathbf{C}$ postmultiplies $\mathbf{B}$. The elements of $\mathbf{A}$ are determined from the following relationship:

$$a_{ij} = \sum_{k=1}^{n} b_{ik}c_{kj} \quad \text{for } i = 1, 2, \ldots, n; \ j = 1, 2, \ldots, n \qquad (A1.3)$$

For example:

$$\begin{bmatrix} 2 & -3 & 1 \\ -5 & 4 & 3 \end{bmatrix} \begin{bmatrix} -6 & 4 & 1 \\ -4 & 2 & 3 \\ 3 & -7 & -1 \end{bmatrix}$$

$$= \begin{bmatrix} 2(-6)+(-3)(-4)+1(3) & 2(4)+(-3)2+1(-7) & 2(1)+(-3)3+1(-1) \\ (-5)(-6)+4(-4)+3(3) & (-5)4+4(2)+3(-7) & (-5)1+4(3)+3(-1) \end{bmatrix}$$

$$= \begin{bmatrix} 3 & -5 & -8 \\ 23 & -33 & 4 \end{bmatrix}$$

Note that the product of a 2 x 3 and a 3 x 3 matrix is a 2 x 3 matrix. Consider four further examples of matrix multiplication:

$$\begin{bmatrix} 1 & 2 \\ 3 & 4 \end{bmatrix}\begin{bmatrix} 5 & 6 \\ 3 & 2 \end{bmatrix} = \begin{bmatrix} 11 & 10 \\ 27 & 26 \end{bmatrix} \qquad \begin{bmatrix} 5 & 6 \\ 3 & 2 \end{bmatrix}\begin{bmatrix} 1 & 2 \\ 3 & 4 \end{bmatrix} = \begin{bmatrix} 23 & 34 \\ 9 & 14 \end{bmatrix}$$

$$\begin{bmatrix} 1 & 2 & 3 \end{bmatrix}\begin{bmatrix} -4 \\ 3 \\ 3 \end{bmatrix} = 11 \qquad \begin{bmatrix} -4 \\ 3 \\ 3 \end{bmatrix}\begin{bmatrix} 1 & 2 & 3 \end{bmatrix} = \begin{bmatrix} -4 & -8 & -12 \\ 3 & 6 & 9 \\ 3 & 6 & 9 \end{bmatrix}$$

In the above examples it is seen that whilst the 2 x 2 matrices can be multiplied in either order the product is different. This is an important observation and in general $\mathbf{BC} \neq \mathbf{CB}$. Note also that multiplying a row by a column vector gives a scalar whereas multiplying a column by a row results in a matrix.

Matrix inversion. The inverse of a square matrix $\mathbf{A}$ is written $\mathbf{A}^{-1}$ and is related to $\mathbf{A}$ by

$$\mathbf{A}\,\mathbf{A}^{-1} = \mathbf{A}^{-1}\mathbf{A} = \mathbf{I}$$

The formal definition of $\mathbf{A}^{-1}$ is

$$\mathbf{A}^{-1} = \mathrm{adj}(\mathbf{A})/\det(\mathbf{A}) \tag{A1.4}$$

where $\mathrm{adj}(\mathbf{A})$ is the adjoint of $\mathbf{A}$. The adjoint of $\mathbf{A}$ is given by

$$\mathrm{adj}(\mathbf{A}) = \Delta^{\mathrm{T}}$$

where Δ is a matrix composed of the cofactors Δ_{ij} of $\mathbf{A}$. Using (A1.4) is not an efficient way to compute an inverse.

A1.6 COMPLEX MATRICES

A matrix can have elements that are complex and such a matrix can be expressed in terms of two real matrices. Thus

$$\mathbf{A} = \mathbf{B} + i\mathbf{C} \quad \text{where } i = \sqrt{(-1)}$$

where $\mathbf{A}$ is complex and $\mathbf{B}$ and $\mathbf{C}$ are real matrices. Conjugation of $\mathbf{A}$ is normally denoted by $\mathbf{A}^*$ and is equal to

$$\mathbf{A}^* = \mathbf{B} - i\mathbf{C}$$

Matrix $\mathbf{A}$ can be transposed so that

$$\mathbf{A}^{\mathrm{T}} = \mathbf{B}^{\mathrm{T}} + i(\mathbf{C}^{\mathrm{T}})$$

Matrix **A** can be transposed *and* conjugated at the same time and this is denoted by $\mathbf{A}^H$. Thus

$$\mathbf{A}^H = \mathbf{B}^T - i(\mathbf{C}^T)$$

For example:

$$\mathbf{A} = \begin{bmatrix} 1-i & -2-3\,i & 4i \\ 2 & 1+2i & 7+5i \end{bmatrix} \quad \mathbf{A}^* = \begin{bmatrix} 1+i & -2+3\,i & -4i \\ 2 & 1-2i & 7-5i \end{bmatrix}$$

$$\mathbf{A}^T = \begin{bmatrix} 1-i & 2 \\ -2-3\,i & 1+2i \\ 4i & 7+5i \end{bmatrix} \qquad \mathbf{A}^H = \begin{bmatrix} 1+i & 2 \\ -2+3\,i & 1-2i \\ -4i & 7-5i \end{bmatrix}$$

It is important to note that the MATLAB expression A' gives the conjugation and transposition of A when applied to a complex matrix, i.e. it is equivalent to $\mathbf{A}^H$. However, A.' gives ordinary transposition which corresponds to $\mathbf{A}^T$.

A1.7 MATRIX PROPERTIES

A real square matrix **A** is,

symmetric if: $\mathbf{A}^T = \mathbf{A}$
skew-symmetric if: $\mathbf{A}^T = -\mathbf{A}$
orthogonal if: $\mathbf{A}^T = \mathbf{A}^{-1}$

A complex square matrix $\mathbf{A} = \mathbf{B} + i\mathbf{C}$ is,

Hermitian if : $\mathbf{A}^H = \mathbf{A}$
unitary if: $\mathbf{A}^H = \mathbf{A}^{-1}$

A1.8 SOME MATRIX RELATIONSHIPS

If **P**, **Q** and **R** are matrices such that

$$\mathbf{W} = \mathbf{PQR}$$

then

$$\mathbf{W}^T = \mathbf{R}^T\,\mathbf{Q}^T\,\mathbf{P}^T \qquad\qquad (A1.5)$$

and

$$\mathbf{W}^{-1} = \mathbf{R}^{-1}\,\mathbf{Q}^{-1}\,\mathbf{P}^{-1} \qquad\qquad (A1.6)$$

If **P**, **Q** and **R** are complex then (A1.6) is still valid and (A1.5) becomes

$$\mathbf{W}^H = \mathbf{R}^H\,\mathbf{Q}^H\,\mathbf{P}^H \qquad\qquad (A1.7)$$

A1.9 EIGENVALUES

Consider the eigenvalue problem

$$\mathbf{Ax} = \lambda \mathbf{x}$$

If $\mathbf{A}$ is an $n \times n$ symmetric matrix then there are n real eigenvalues, λ_i, and n real eigenvectors, $\mathbf{x}_i$, that satisfy this equation. If $\mathbf{A}$ is an $n \times n$ Hermitian matrix then there are n real eigenvalues, λ_i, and n complex eigenvectors, $\mathbf{x}_i$, that satisfy the eigenvalue problem. The polynomial in λ given by $\det(\mathbf{A} - \lambda \mathbf{I}) = 0$ is called the characteristic equation. The roots of this polynomial are the eigenvalues of $\mathbf{A}$. The sum of the eigenvalues of $\mathbf{A}$ equals trace($\mathbf{A}$) where trace($\mathbf{A}$) is defined as the sum of the elements on the leading diagonal of $\mathbf{A}$. The product of the eigenvalues of $\mathbf{A}$ equals $\det(\mathbf{A})$.

It is interesting to note that if we define $\mathbf{C}$ as

$$\mathbf{C} = \begin{bmatrix} -p_1/p_0 & -p_2/p_0 & \cdots & -p_{n-1}/p_0 & -p_n/p_0 \\ 1 & 0 & \cdots & 0 & 0 \\ 0 & 1 & \cdots & 0 & 0 \\ \vdots & \vdots & & \vdots & \vdots \\ 0 & 0 & \cdots & 1 & 0 \end{bmatrix}$$

then the eigenvalues of $\mathbf{C}$ are the roots of the polynomial

$$p_0 x^n + p_1 x^{n-1} + \ldots + p_{n-1} x + p_n = 0$$

The matrix $\mathbf{C}$ is called the companion matrix.

A1.10 DEFINITION OF NORMS

The p-norm for the vector $\mathbf{v}$ is defined thus:

$$\|\mathbf{v}\|_p = (|v_1|^p + |v_2|^p + \ldots + |v_n|^p)^{1/p} \tag{A1.8}$$

The parameter p can take any value but only three values are commonly used. If $p = 1$ in (A1.8) we have the 1-norm, $\|\mathbf{v}\|_1$, thus:

$$\|\mathbf{v}\|_1 = |v_1| + |v_2| + \ldots + |v_n| \tag{A1.9}$$

If $p = 2$ in (A1.8) we have the 2-norm or Euclidean norm of the vector $\mathbf{v}$ which is written $\|\mathbf{v}\|$ or $\|\mathbf{v}\|_2$ and is defined thus:

$$\|\mathbf{v}\|_2 = \sqrt{v_1^2 + v_2^2 + \ldots + v_n^2} \tag{A1.10}$$

Note that it is not necessary to take the modulus of the elements because in this case each element value is squared. The Euclidean norm is also called the length of the vector.

These names arise from the fact that in two- or three-dimensional Euclidean space a vector of two or three elements is used to specify a position in space. The distance from the origin to the specified position is identical to the Euclidean norm of the vector.

If p tends to infinity in (A1.8) we have $\|v\|_\infty = \max(|v_1|, |v_2|, ..., |v_n|)$, the infinity norm. At first sight this might appear inconsistent with (A1.8). However, when p tends to infinity the modulus of each element is raised to a very large power and the largest element will dominate the summation.

These functions are implemented in MATLAB; `norm(v,1)`, `norm(v,2)` (or `norm(v)`) and `norm(v,inf)` return the 1, 2 and infinity norms of the vector v, respectively.

A1.11 REDUCED ROW ECHELON FORM

The RREF of a matrix also has an important role to play in the theoretical understanding of linear algebra. A matrix is transformed into its RREF when the following conditions have been met:

(1) all zero rows, if they exist, are at the bottom of the matrix,

(2) the first non-zero element in every non-zero row is unity,

(3) for each non-zero row, the first non-zero element appears to the right of the first non-zero element of the preceding row,

(4) for any column in which the first non-zero element of a row appears, all other elements are zero.

The RREF is determined by using a finite sequence of elementary row operations. The RREF is a standard form and it is the most fundamental form of a matrix that can be achieved using elementary row operations alone.

For a system of equations $\mathbf{Ax} = \mathbf{b}$ we can define the augmented matrix $[\mathbf{A}\ \mathbf{b}]$. If this matrix is transformed to its RREF, the following may be deduced:

(1) If $[\mathbf{A}\ \mathbf{b}]$ is derived from an inconsistent system (i.e. no solution exists) the RREF has a row of the form $[0\ ...\ 0\ 1]$.

(2) If $[\mathbf{A}\ \mathbf{b}]$ is derived from a consistent system with an infinity of solutions then the number of columns of the coefficient matrix is greater than the number of non-zero rows in the RREF, otherwise there is a unique solution and it appears in the last (augmented) column of the RREF.

(3) A zero row in the RREF indicates that the original set of equations contained equations with redundant information, i.e. information contained in other equations of the system.

In computing the RREF, numerical problems can arise that are common to other procedures that use elementary row operations, see section 2.6.

Appendix 2 – List of MATLAB functions

This appendix provides a list of the major MATLAB functions we have developed. The name of the function is provided, its purpose and the page on which it is defined. These functions can be obtained on disk from The Mathworks, Inc.

References

Abramowitz, M., and Stegun, I.A., (1965). *Handbook of Mathematical Functions*. 9th Edition, Dover, New York.

Adby, P.R., and Dempster, M.A.H., (1974). *Introduction to Optimisation Methods*. Chapman and Hall, London.

Armstrong, R., and Kulesza, B.L.J., (1981). "An approximate solution to the equation $x = \exp(-x/e)$". *Bulletin of the Institute of Mathematics and its Applications*, **17** (2-3) 56.

Barnes, E.R., (1986). "Affine transform method". *Mathematical Programming*, **36** 174-182.

Beltrami, E.J., (1987). *Mathematics for Dynamic Modelling*. Academic Press, Boston.

Brent, R.P., (1971). "An algorithm with guaranteed convergence for finding the zero of a function". *Computer Journal*, **14** 422-425.

Brigham, E.O., (1974). *The Fast Fourier Transform*. Prentice Hall, Englewood Cliffs, N. J.

Butcher, J.C., (1964). "On Runge Kutta processes of high order". *Journal of the Australian Mathematical Society*, **4** 179-194.

Cooley, P.M., and Tukey, J.W., (1965). "An algorithm for the machine calculation of complex Fourier series". *Mathematics of Computation*, **19** 297-301.

Dantzig, G.B., (1963). *Linear Programming and Extensions*. Princeton University Press.

Dekker, T.J., (1969). "Finding a zero by means of successive linear interpolation" in Dejon, B. and Henrici, P. (Eds.). *Constructive Aspects of the Fundamental Theorem of Algebra*. Wiley-Interscience, New York.

Dongarra, J.J., Bunch, J., Moler, C.B., and Stewart, G., (1979). *LINPACK User's Guide*. SIAM, Philadelphia.

Dowell, M., and Jarrett, P., (1971). "A modified *regula falsi* method for computing the root of an equation". *BIT*, **11** 168-174.

Fletcher, R., and Reeves, C.M., (1964). "Function minimisation by conjugate gradients". *Computer Journal,* **7** 149-154.

Fox, L., and Mayers, D.F., (1968). *Computing Methods for Scientists and Engineers.* Oxford University Press.

Froberg, C.-E., (1969). *Introduction to Numerical Analysis.* 2nd Edition. Addison-Wesley, Reading, Mass.

Garbow, B.S., Boyle, J.M., Dongarra, J.J., and Moler, C.B., (1977). *Matrix Eigensystem Routines: EISPACK Guide Extension,* Lecture Notes in Computer Science. **51** Springer Verlag, Berlin.

Gear, C.W., (1971). *Numerical Initial Value Problems in Ordinary Differential Equations.* Prentice Hall, Englewood Cliffs, N. J.

Gilbert, J.R., Moler, C.B., and Schreiber, R., (1992). "Sparse matrices in MATLAB: design and implementation". *SIAM Journal of Matrix Analysis and Application,* **13** (1) 333-356.

Gill, S., (1951). "Process for the step by step integration of differential equations in an automatic digital computing machine". *Proceedings of the Cambridge Philosophical Society,* **47** 96-108.

Goldberg, D.E., (1989). *Genetic Algorithms in Search, Optimisation and Machine Learning.* Addison-Wesley, Reading, Mass.

Golub, G.H., and Van Loan, C.F., (1989). *Matrix Computations.* 2nd Edition, John Hopkins University Press.

Gragg, W.B., (1965). "On extrapolation algorithms for ordinary initial value problems". *SIAM Journal of Numerical Analysis,* **2** 384-403.

Guyan, R.J., (1965). "Reduction of stiffness and mass matrices". *AIAA Journal,* **3** (2) 380.

Hamming, R.W., (1959). "Stable predictor–corrector methods for ordinary differential equations". *Journal of the ACM,* **6** 37-47.

Hopfield, J.J., and Tank, D.W., (1985). "Neural computation of decisions in optimisation problems". *Biological Cybernetics,* **52** (3) 141-152.

Hopfield, J.J., and Tank, D.W., (1986). "Computing with neural circuits: a model". *Science,* **233** 625-633.

Jeffrey, A., (1979). *Mathematics for Engineers and Scientists.* Nelson, Sunbury-on-Thames.

Karmarkar, N.K., (1984). "A new polynomial time algorithm for linear programming". *AT & T Bell Laboratories, Murray Hill, N. J.*

Karmarkar, N.K., and Ramakrishnan, K.G., (1991). "Computational results of an interior point algorithm for large scale linear programming". *Mathematical Programming,* **52** (3) 555-586.

Lambert, J.D., (1973). *Computational Methods in Ordinary Differential Equations.* John Wiley & Sons, London.

Lindfield, G.R., and Penny, J.E.T., (1989). *Microcomputers in Numerical Analysis.* Ellis Horwood, Chichester.

MATLAB User's Guide. (1989). The MathWorks, Inc. [This describes an earlier version of MATLAB]

Merson, R.H., (1957). "An operational method for the study of integration processes". *Proc. Conf. on Data Processing and Automatic Computing Machines. Weapons Research Establishment. Salisbury, South Australia.*

Press, W.H., Flannery, B.P., Teukolsky, S.A., and Vetterling, W.T., (1990). *Numerical Recipes: The Art of Scientific Computing in Pascal.* Cambridge University Press.

Ralston, A., (1962). "Runge Kutta methods with minimum error bounds". *Mathematics of Computation,* **16** 431-437.

Ralston, A., and Rabinowitz, P., (1978). *A First Course in Numerical Analysis.* McGraw-Hill, New York.

Ramirez, R.W., (1985). *The FFT, Fundamentals and Concepts.* Prentice Hall, Englewood Cliffs, N. J.

Salvadori, M.G., and Baron, M.L., (1961). *Numerical Methods in Engineering.* Prentice Hall. London.

Short, L., (1992). "Simple iteration behaving chaotically". *Bulletin of the Institute of Mathematics and its Applications,* **28** (6-8) 118-119.

Simmons, G.F., (1972). *Differential Equations with Applications and Historical Notes.* McGraw-Hill. New York.

Smith, B.T., Boyle, J.M., Dongarra, J.J., Garbow, B.S., Ikebe, Y., Kleme, V.C., and Moler, C., (1976). *Matrix Eigensystem Routines: EISPACK Guide.* Lecture Notes in Computer Science, **6** 2nd Edition, Springer Verlag. Berlin.

Styblinski, M.A., and Tang, T.-S., (1990). "Experiments in non-convex optimisation: stochastic approximation with function smoothing and simulated annealing". *Neural Networks,* **3** (4) 467-483.

Solutions to problems

Note that scripts may be compressed to save space.

Chapter 1

1.1. (a) Since some x are negative the corresponding square roots are imaginary and i is used. (b) In executing x./y the divide by zero produces the symbol ∞ and a warning.

1.2. (b) Note that t2 is identical to c but t1 is not since the sqrt function gives the square root of the individual elements of c.

1.4. $x = 2.4545, y = 1.4545, z = -0.2727$. Note when using the / operator the solution is given by: x=b'/a'.

1.8. The plot does not truly represent the function $\cos(x^3)$ because there are insufficient plotting points.

1.9. Function fplot automatically adjusts to provide a smoother plot. However, changing x to -2:.01:2 gives a similar quality graph using the function plot.

1.12. $x = 1.6180$.

1.14. Using $x_1 = 1, x_2 = 2, ..., x_6 = 6$, a possible script is

```
n=6; x=1:n; for j=1:n, p(j)=1; for i=1:n, if i~=j, p(j)=p(j)*x(i);
end; end; end; p
```

1.15. A suitable script is

```
x=.82; tol=.005; s=x; i=2; term=x;
while abs(term)>tol
   term=-term*x; s=s+term/i; i=i+1;
end
s, log(1+x)
```

1.17. A possible form of the function is

```
function [x1,x2]=funct1(a,b,c)
d=b*b-4*a*c;
if d==0
  x1=-b/(2*a); x2=x1;
else
  x1=(-b+sqrt(d))/(2*a); x2=(-b-sqrt(d))/(2*a);
end
```

1.18. A possible script is

```
function [x1,x2]=funct2(a,b,c)
if a ~= 0
  % as in problem 1.17
else
  disp('warning only one root'); x1=-c/b; x2=x1;
end
```

1.19. The graph provides an initial approximation of 1.5. Use the function call
       fzero('funct3',1.5) to obtain the root as 1.2512.

Chapter 2

2.1. Exact solution is $x = [-12.5 \ -24 \ -34 \ -42 \ -47.5 \ -50 \ -49 \ -44 \ -34.5 \ -20]^T$.
      The Gauss–Seidel method requires 149 iterations and the Jacobi method
      requires 283 iterations to give the result to the required accuracy.

2.2.
```
n   norm(p-r)     norm(q-r)
3   0.0000        0.0000
4   0.0849        0.0000
5   84.1182       0.1473
6   4.7405e10     6.7767e3
```

Note the large error in the inverse of the square of the Hilbert matrix when $n = 6$.

2.3. For $n = 3$, 4, 5 and 6, the answer is 2.7464E05, 2.4068E08, 2.2715E11 and
      2.2341E14 respectively. The large errors in problem 2.2 arise from the fact that
      the Hilbert matrix is very ill-conditioned, as shown by these results.

2.4. For example, with $n = 5$; $a = 0.2$; $b = 0.1$; thus $a + 2b < 1$ and maximum error
      in the matrix coefficients is 1.0412E–05. Taking $n = 5$; $a = 0.3$; $b = 0.5$; thus
      $a + 2b > 1$ and after 10 terms maximum error in matrix coefficients is 10.8770.
      After 20 terms maximum error is 50.5327, clearly diverging.

2.5.   Eigenvalues are 5, $2 + 2i$ and $2 - 2i$. Thus taking $\lambda = 5$ in the matrix $(\mathbf{A} - \lambda\mathbf{I})$ and finding the RREF gives

$$\mathbf{p} = \begin{bmatrix} 1 & 0 & -1.3529 \\ 0 & 1 & 0.6471 \\ 0 & 0 & 0 \end{bmatrix}$$

Hence $\mathbf{px} = \mathbf{0}$. Solving this gives $x_1 = 1.3529x_3$, $x_2 = -0.6471x_3$ and x_3 is arbitrary.

2.6.

n	flops (full)	flops(sparse)
20	4488	408
30	13073	638
40	28558	850
50	52943	1089

2.7.

n	flops (full)	flops(sparse)
10	2206	630
20	13780	2014
30	43458	4902

2.8.   Flops (full) = 403, flops (sparse) = 50.

2.9. $\mathbf{x}^T = [0.9500\ 0.9811\ 0.9727]$. All methods give identical solution. Note if `[q,r]=qr(a)` and `y=q'*b`; then `x=r(1:3,1:3)\y(1:3)`.

2.10. The solution is $[0\ 0\ 0\ 0 \ldots n + 1]$.

2.13. For $n = 20$ condition number is 178.0643; theoretical condition number is 162.1139. For $n = 50$ condition number is 1053.5; theoretical condition number is 1013.2.

2.14. Right-hand vectors are

$$\begin{bmatrix} 0.0484 + 0.4446i \\ -0.3962 + 0.4930i \\ 0.4930 + 0.3962i \end{bmatrix} \quad \begin{bmatrix} 0.0484 - 0.4446i \\ -0.3962 - 0.4930i \\ 0.4930 - 0.3962i \end{bmatrix} \quad \begin{bmatrix} 0.4082 \\ 0.8165 \\ 0.4082 \end{bmatrix}$$

The corresponding eigenvalues are $2 + 4i$, $2 - 4i$ and 1. The left-hand vectors are obtained by using the function `eig` on the transposed matrix.

Chapter 3

3.2.   The solution is 27.8235.

3.3.   The solutions are -2 and 1.6344.

3.4. For $c = 5$, with the initial approximation 1.3 or 1.4, the root 1.3735 is obtained after two or three iterations. When $c = 10$, with the initial approximation 1.4, the root 1.4711 is obtained after five iterations. With initial approximation 1.3, convergence is to 193.1083 after 41 iterations. This is a root but the discontinuity in the function has degraded the performance of the Newton algorithm.

3.5. Schroder's method provides the solution $x = 1$ in one iteration but Newton's method gives $x = 1.0011$ and requires 87 iterations. Since the solution obtained by Schroder's method is very accurate the denominator becomes very close to zero and a divide by zero warning is displayed.

3.6. The equation can be rearranged into the form $x = \exp(x/10)$. Iteration gives $x = 1.1183$. There may be other successful rearrangements.

3.7. The solution is $E = 0.1280$.

3.8. The answers are -3.019×10^{-6} and -6.707×10^{-6} for initial values 1 and -1.5 respectively. The exact solution is clearly 0 but this is a difficult problem.

3.9. The three answers are 1.4299, 1.4468 and 1.4458 which are obtained for four, five and six terms respectively. These answers are converging to the correct answer.

3.10. Both approaches give identical results, $x = 8.2183$, $y = 2.2747$. The single variable function is $x/5 - \cos x = 2$. Alternatively the following call can be used

```
newtonmv([1 1]','p318','p319',2,1e-4)
```

It requires the functions and derivatives to be defined thus

```
function v=p318(x)
v=zeros(2,1); v(1)=exp(x(1)/10)-x(2);
v(2)=2*log(x(2))-cos(x(1))-2;
```

```
function vd=p319(x)
vd=zeros(2,2); vd(1,:)=[exp(x(1)/10)/10 -1];
vd(2,:)=[sin(x(1)) 2/x(2)];
```

3.11. The solution given by broyden is $x = 0.1605$, $y = 0.4931$.

3.12. A solution is $x = 0.9397$, $y = 0.3420$. The MATLAB function newtonmv requires seven iterations and broyden requires 33.

3.14. The five roots are 1, $-i$, i, $-\sqrt{2}$, $\sqrt{2}$.

3.15. Solution is $x = -0.1737$, $-0.9848i$, $0.9397 + 0.3420i$ and $-0.7660 + 0.6428i$. This is identical to the exact answer.

3.16. The MATLAB function required is

```
function v=jarrett(f,x1,x2,tol)
gamma=.5;d=1;
while abs(d) >tol
  f2=feval(f,x2);f1=feval(f,x1);
  df=(f2 -f1)/(x2-x1); x3=x2-f2/df; d=x2-x3;
  if f1*f2 >0
    x2=x1; f2=gamma*f1;
  end
  x2=x3
end
```

3.17. The third-order method provides the required accuracy after seven iterations. The second-order method requires 10 iterations.

3.18. The graphs show that for $c = 2.8$ there is convergence to a single solution, for $c = 3.25$ the iteration oscillates between two values, for $c = 3.5$ the iteration oscillates between four values and for $c = 3.8$ there is chaotic oscillation between many values.

Chapter 4

4.1. First derivative is 0.2391, the second derivative is –2.8256. The function diffgen gives accurate answers using either $h = 0.1$ or 0.01. The function changes slowly over this range of values.

4.2. When $x = 1$ the computed and exact derivative is –5.0488, when $x = 2$ the computed derivative is –176.6375 (exact = –176.6450) and when $x = 3$ the computed derivative is –194.4680 (exact = –218.6079).

4.3. Using the new formula for problem 4.1, the first derivative estimate is 0.2267 and 0.2390 for $h = 0.1$ and 0.01 respectively. The second derivative is –2.8249 and –2.8256 for $h = 0.1$ and 0.01 respectively. In problem 4.2 for $x = 1$, 2 and 3 the first derivative estimates are –5.0489, –175.5798 and –150.1775 respectively. Note that these are less accurate than using diffgen.

4.4. The approximate derivatives are –1367.2, –979.4472, –1287.7 and –194.4680. If h is decreased to 0.0001 then the values are the same as the exact derivatives to the given number of decimal places.

4.5. The exact partial derivatives are 593.652 and 445.2395 with respect to x and y respectively. The corresponding approximate values are 593.7071 and 445.2933.

4.6. The integral method estimates 6.3470 primes in the range 1 to 10, 9.633 primes in the range 1 to 17 and 15.1851 primes in the range 1 to 30. The actual numbers are 7, 10 and 15.

4.7.   Exact values are 1.5708, 0.5236 and 0.1428 for $r = 0$, 1 and 2 respectively. Approximations provided by integral are 1.5338, 0.5820 and 0.2700.

4.8.   The exact values are -0.0811 for $a = 1$ and 0.3052 for $a = 2$. Using `simp1` with 512 points gives agreement to 12 decimal places.

4.9.   The exact answer is -0.915965591 and the answer given by `fgauss` is -0.9136. Function `simp1` cannot be used because of the singularity at $x = 0$.

4.10.   The exact answer is 0.915965591, `fgauss` gives 0.9159655938. Note that the integrals of 4.9 and 4.10 have the same value apart from the sign.

4.11.   (i) Using (4.9.1) with 10 points gives 3.97746326050642, 16 point Gauss gives 3.8145. (ii) Using (4.9.2) gives 1.77549968921218, 16 point Gauss gives 1.7758.

4.13.   Filon gives
        2.00000000000098, -0.13333333344440 and $-2.000199980281494E{-}04$.

4.14.   Using Romberg's method with nine divisions gives $-2.000222004003794E{-}4$. Using Simpson's rule with 1024 intervals gives $-1.999899106566088E{-}4$.

4.17.   The solution for (i) is 48.96321182552904 and for (ii) 9.726564917628732E03. These compare well with the exact solution which can be computed from the formula $4\pi^{(n+1)}/(n + 1)^2$ where n is the power of x and y.

4.18.   (i) Fix limits by substituting $y = \{\sqrt{(x/3)} - 1\}z + 1$. Answer: -1.71962748468952. (ii) To fix limits substitute $y = (2 - x)z$. Answer: 0.22222388780205.

4.19.   The answers are (i) -1.71821293254848, (ii) 0.22222222200993.

4.20.   Values of the integral are given in the following table:

z	exact	16 point Gauss
0.5	0.493107418	0.49310741784618
1	0.946083070	0.94608306999140
2	1.605412977	1.60541297617644

Chapter 5

5.1.   When $t = 10$ the exact value is 30.326533.
        `feuler` 29.9368, 30.2885, 30.3227 with $h = 1$, 0.1 and 0.01 respectively.
        `eulertp` 30.3281, 30.3266 with $h = 1$ and 0.1 respectively.
        `rkgen`   30.3265 with $h = 1$.

5.2.   Classical method gives 108.9077, Butcher method gives 109.1924, Merson method gives 109.0706. Exact answer is $2 \exp(x^2) = 109.1963$.

5.3. Adams–Bashforth–Moulton method gives 4.1042, Hamming's method gives 4.1043. Exact answer is 4.1042499.

5.4. Using ode23 gives 0.0456, using ode45 gives 0.0588.

5.5. Solution of problem 5.1 with $h = 1$ is 30.3265. Solution of problem 5.2 with $h = 0.2$ is 108.8906. Solution of problem 5.2 with $h = 0.02$ is 109.1963.

5.6. (i) 7998.6, exact = 8000. (ii) 109.1963.

5.9. Method is stable for $h = 0.1$ and 0.2, and unstable for $h = 0.4$.

Chapter 6. *NB. FD means finite difference.*

6.1. (i) hyperbolic, (ii) parabolic, (iii) $\phi(x, y) > 0$, hyperbolic, $\phi(x, y) < 0$ elliptic.

6.2. Initial slope $= -1.6714$. Shooting and FD methods give good results.

6.3. This is an example of a stiff equation. (i) The actual slope when $x = 0$ is 1.0158×10^{-24}. Because we cannot determine this slope accurately the shooting method gives a very inaccurate solution. (ii) In this case the shooting method provides a good result because the initial slope is -120. In both cases the FD method requires a large number of divisions to give an accurate result.

6.5. Finite difference method gives $\lambda_1 = 2.4623$. Exact $\lambda_1 = (\pi/L)^2 = 2.4674$.

6.6. At $t = 0.5$ the variation of z is almost linear between the boundaries at 0 and 10.

6.7. The exact and FD approximations are very similar.

6.8. The exact and FD approximations are similar with a maximum error of 0.0479.

6.9. $\lambda = 5.8870, 14.0418, 19.6215, 27.8876, 29.8780$.

6.10. [0.7703 1.0813 1.5548 1.583 1.1943 1.5548 1.583 1.194 1.0813 0.7703].

Chapter 7

7.1. Using the aitken function, $E(2°) = 1.5703$, $E(13°) = 1.5507$, $E(27°) = 1.4864$. These are accurate to the places given.

7.2. The root is 27.8235.

7.3. (i) $p(x) = 0.9814x^2 + 0.1529$ and $p(x) = -1.2083x^4 + 2.1897x^2 + 0.0137$. The fourth degree polynomial gives a good fit.

7.4. Interpolation gives 0.9284 (linear), 0.9463 (spline), 0.9429 (cubic polynomial). The MATLAB function aitken gives 0.9455. This is the exact value to four decimal places.

7.5. $p(x) = -0.3238x^5 + 3.2x^4 - 6.9905x^3 - 12.8x^2 + 31.1429x$. Note that the polynomial oscillates between data points. The spline does not exhibit this characteristic, suggesting that it better represents any underlying function from which the data might have been taken.

7.6.   (i) $f(x) = 3.1276 + 1.9811e^x + e^{2x}$.
       (ii) $f(x) = 685.1 - 2072.2/(1 + x) + 1443.8/(1 + x)^2$.
       (iii) $f(x) = 47.3747x^3 - 128.3479x^2 + 103.4153x - 5.2803$.
       Plotting these functions shows that the best fit is given by (i). The polynomial fit is a reasonable one.

7.8.   Product of primes less than P given by $0.3679 + 1.0182 \log_e P$ approximately.

7.9. $a_0 = 1$, $a_1 = -0.5740$, $a_2 = 0.9456$, $a_3 = -0.6865$, $a_4 = 0.4115$, $a_5 = -0.0966$.

7.10. Exact: -78.3323. Interpolation gives -78.3340 (cubic) or -77.9876 (linear).

7.11. The plot should display a smooth airfoil section.

7.12. The data is sampled from $y = \sin(2\pi f_1 t) + 2\cos(2\pi f_2 t)$ where $f_1 = 1.25$ Hz and $f_1 = 3.4375$ Hz. At 1.25 Hz, DFT $= 15.9999i$ and 3.4375 Hz, DFT $= 32.0001$. Note that to relate the size of the DFT components to the frequency components in the data we divide the DFT by the number of samples and multiply by 2.

7.13. Algebraically, $32 \sin^5(30t) = 20 \sin(30t) - 10 \sin(90t) + 2 \sin(150t)$ and $32 \sin^6(30t) = 10 - 15 \sin(60t) + 6 \cos(120t) - \cos(180t)$. To verify these results from the DFT it is necessary to divide it by n and multiply by 2. Note also that the coefficient at zero frequency is 20, not 10. This is a consequence of the definition of the DFT, see page 248.

7.14. Components in spectrum at 30 Hz and 112 Hz. The reason for the large component at 112 Hz is that the component in the data at 400 Hz is above the Nyquist frequency and is folded back to give a spurious component, i.e. 400 Hz is 144 Hz above the Nyquist frequency of 256 Hz; 112 Hz is 144 Hz below it.

7.16. With 32 points the frequency increment is 16 Hz and the significant components are at 96 Hz and 112 Hz (the largest amplitude). With 512 points the frequency increment is reduced to 1 Hz and the significant components are at 103, 104 and 105 Hz with the largest amplitude at 104 Hz. The original data had a frequency component of 104.5 Hz.

Chapter 8

8.1.   Objective is 21.6667. Solution is $x_1 = 3.6667$, $x_3 = 0.3333$, other variables are zero.

8.2.   Objective is –21.6667. Solution is $x_1 = 3.3333$, $x_2 = 1.6667$, $x_4 = 0.3333$, other variables are zero. Thus this problem and the previous one have objective functions of equal magnitude.

8.3.   Objective is –100. Solution is $x_1 = 10$, $x_3 = 20$, $x_5 = 22$, other variables are zero.

8.4.   This is a difficult function for the conjugate gradient method and this is why the accuracy of the line-search for the built-in MATLAB function fmin was changed to produce more accurate results. Solution is [1.0007 1.0014] with gradient 1.0E–03 x [0.3386 0.5226].

8.5.   Exact and computed solutions are both [–2.9035 –2.9035 1 1 1].

8.6.   Solution is [–0.4600 0.5400 0.3200 0.8200]^T. Norm($\mathbf{b} - \mathbf{A}\mathbf{x}$) = 1.3131E–14.

8.7.   [xval,maxf]=optga('p802',[0 2],8,12,20,.005,.6) where p802 is a MATLAB function defining the problem. A test run gave the following answers: xval = 0.9098, maxf = 0.4980.

8.8.   The major modification is to the fitness function as follows:

```
function [fit,fitot]=fitness2d(criteria,chrom,a,b)
%calculate fitness of a set of chromosomes for a two variable
%function assuming each variable is defined in the range a to
%b using a two variable function given by criteria
[pop bitl]=size(chrom); vlength=floor(bitl/2);
for k=1:pop
  v=[ ];v1=[ ];v2=[ ]; partchrom1=chrom(k,1:vlength);
  partchrom2=chrom(k,vlength+1:2*vlength);
  v1=binvreal(partchrom1,a,b); v2=binvreal(partchrom2,a,b);
  v=[v1 v2]; fit(k)=feval(criteria,v);
end;
fitot=sum(fit);
```

The given function fun must be defined as a two-variable function. A call of the modified algorithm is optga2d('f80n',[1 2],24,40,100,.005,.6), for example, where f80n defines $z = x^2 + y^2$. This gives the sample results maxf=7.9795 and xval=[1.9956 1.9993].

Index